TEACHING AS ACTIVISM

Teaching as Activism

Equity Meets Environmentalism

EDITED BY
PEGGY TRIPP AND LINDA MUZZIN

McGill-Queen's University Press
Montreal & Kingston • London • Ithaca

© McGill-Queen's University Press 2005
ISBN 0-7735-2807-5 (cloth)
ISBN 0-7735-2808-3 (paper)

Legal deposit fourth quarter 2005
Bibliothèque nationale du Québec

Printed in Canada on acid-free paper that is 100% ancient forest free (100% post-consumer recycled, processed chlorine free).

This book has been published with the help of a grant from the Canadian Federation for the Humanities and Social Sciences, through the Aid to Scholarly Publications Programme, using funds provided by the Social Sciences and Humanities Research Council of Canada.

McGill-Queen's University Press acknowledges the support of the Canada Council for the Arts for our publishing program. We also acknowledge the financial support of the Government of Canada through the Book Publishing Industry Development Program (BPIDP) for our publishing activities.

Time Capsule, page 180, copyright Linda Stitt

National Library of Canada Cataloguing in Publication

Teaching as activism : equity meets environmentalism / edited by Peggy Tripp and Linda Muzzin.

Includes bibliographical references and index.
ISBN 0-7735-2807-5 (bound). – ISBN 0-7735-2808-3 (pbk.)

1. Teaching. 2. Social ethics – Study and teaching.
3. Environmental responsibility – Study and teaching.
4. Science – Social aspects – Study and teaching. 5. Humane education. 6. Adult education. 7. Continuing education.
8. Education, Higher. I. Tripp, Peggy II. Muzzin, Linda June, 1948–

LC191.T42 2005   374′.1102   c2004-904803-1

This book was typeset by True to Type in 10/13 Palatino

For Wayne

Contents

Figures xi

Acknowledgments xiii

Contributors xv

Prologue
Ursula Franklin xxi

Overview
Peggy Tripp and Linda Muzzin 3

SECTION ONE PERSONAL EXPERIENCES AND PARADIGM SHIFTS

1 The Seedlings Mattered
   Heather Menzies 25

2 "The Wolf Must Not Be Made a Fool Of": Reflections on Education, Ethics, and Epistemology
   Bob Jickling 35

3 Hoops of Spirituality in Science and Technology
   Njoki Wane and Barbara Waterfall 47

4 Teaching Sustainable Science
   Peggy Tripp 65

5 Professional Ideology and Educational Practice: Learning to Be a Health Professional
Moira Grant 80

6 Mainstreaming Transformative Teaching
Ann Matthews 95

7 Science, Environment, and Women's Lives: Integrating Teaching and Research
Marianne Ainley 107

SECTION TWO PROBLEMATIZATION OF DOMINANT REALITIES

8 You Can't Be the Global Doctor If You're the Colonial Disease
Marie Battiste 121

9 Colonialism and Capitalism: Continuities and Variations in Strategies of Domination and Oppression
Vanaja Dhruvarajan 134

10 The Brave New World of Professional Education
Linda Muzzin 149

11 Working in the Field of Biotechnology
Elisabeth Abergel 167

SECTION THREE WEAVING NEW WORLDS AND RECLAIMING SUBJUGATED KNOWLEDGES

12 The Illiteracy of Social Scientists with Respect to Environmental Sustainability
Margrit Eichler 181

13 Teaching in Engineering As If the World Mattered
Monique Frize 195

14 Evaluation Matters: Creating Caring 'Rules' in the Human Science Paradigm in Nursing Education
Alexandra McGregor 210

15 Post-colonial Remedies for Preserving Indigenous Knowledge and Heritage
Marie Battiste 224

16 The Anishinaabe Teaching Wand and Holistic Education
   Robin Cavanagh   233

17 Visions for Embodiment in Technoscience
   Natasha Myers   255

18 Bioregional Teaching: How to Climb, Eat, Fall, and Learn from Porcupines
   Leesa Fawcett   269

   Index   281

Figures

Spring ferns   *Lori Fox Rossi*   2
Wolf and hand print   *Lori Fox Rossi*   34
Turtle Island   *Eileen Antone*   46
Boreal trees   *Karen Helmer*   64
Barrow's goldeneye duck   *Marianne Ainley*   106
Colonial Marinade   *Diane Rae*   120
The Anishinaabe Teaching Wands (chapter 16 figures)   *Christine Vogel* © Robin Nimkii Cavanagh 2000   237–49
Entangled vines   *Natasha Myers*   254
Porcupine   *Lori Fox Rossi*   268

Acknowledgments

We would like to thank the Social Sciences and Humanities Research Council of Canada from which we received funding in the Women and Change competition for the SSHRC Strategic Network, Biology As If The World Mattered (BAITWORM). BAITWORM was important in bringing the ideas in this book together, and we would especially like to acknowledge the contributions of BAITWORM members Marilyn Macdonald, Jamie Lynn Magnusson, Myron Frankman, Karen Messing, Daniel Hollenberg, Deboleena Roy, David Phillips, Begna Dugassa, and Peter Pennefather. We enjoyed working with Imogen Brian and Joan McGilvray of the Press. Special thanks to Tina Martimianakis for her assistance with the index. We would also like to thank the Aid to Scholarly Publishing program officers, who provided helpful suggestions, and finally, the anonymous reviewers of the book.

Contributors

AUTHORS

ELISABETH ABERGEL is an assistant professor in the International Studies Program in the Multidisciplinary Studies Department at Glendon College, York University. With a background in molecular genetics, she now teaches courses on science and technology, international relations, and globalization; her research focuses on the international governance of technoscience.

MARIANNE GOSZTONYI AINLEY is a historian of science, naturalist, and artist. Although she retired from teaching at the University of Northern British Columbia (UNBC), she is an adjunct professor of History (UNBC) and of Women's Studies (University of Victoria), is completing two books on Canadian women and science; and is working on a comparative interdisciplinary project on gender, environments, and the transfer of knowledge in nineteenth- and twentieth-century Canada and Australia.

MARIE BATTISTE is Mi'kmaq, from the Potlo'tek First Nation in Cape Breton, Nova Scotia. She is professor in research and leadership in Aboriginal Education, formerly Indian and Northern Education Program, in the Department of Educational Foundations at the University of Saskatchewan (since 1993).

ROBIN NIMKII CAVANAGH is an Anishinaabe helper from northern Ontario. He is currently working on his dissertation, "The Development of an Anishinaabe Spiritual Educational Pedagogy," in the Faculty of Environmental Studies at York University.

VANAJA DHRUVARAJAN, professor/senior scholar at the University of Winnipeg, is serving as adjunct professor in the Pauline Jewett Institute of Women's Studies in Ottawa. Her latest publication is the co-authored book *Gender, Race and Nation: A Global Perspective*, published by the University of Toronto Press.

MARGRIT EICHLER is professor of Sociology and Equity Studies at the Ontario Institute for Studies in Education of the University of Toronto and a fellow of the Royal Society of Canada. Formerly the director of the University of Toronto's Institute for Gender Studies and Women's Studies, she is a sociologist who applies a feminist perspective to family studies, eco-sociology, and research methods and a historical approach to the development of Women's Studies in Canada.

LEESA FAWCETT is associate professor and PhD Program Coordinator in the Faculty of Environmental Studies at York University in Toronto. Her intellectual passions encompass a feminist and environmental philosophy perspective on animal studies, environmental education, and conservation.

URSULA FRANKLIN is an activist and experimental physicist, University Professor Emerita at the University of Toronto, a former board member of the National Research Council and the Science Council of Canada, an officer of the Order of Canada, and a fellow of the Royal Society of Canada.

MONIQUE FRIZE is a professor in systems and computer engineering at Carleton University and in the School of Engineering and Information Technology at the University of Ottawa. Her research interests are: biomedical engineering, ethics for engineers, and women in science and engineering. She is president of the International Network of Women Engineers and Scientists.

MOIRA GRANT is program co-ordinator of the medical laboratory science program at the University of Ontario Institute of Technology in

Contributors xvii

Oshawa, Ontario. True to her poststructural leanings, she negotiates a scenic path that weaves through her interests in medical laboratory science, professional education, feminist science critique, and health policy.

BOB JICKLING, a long-time Yukon resident and founding editor of the *Canadian Journal of Environmental Education*, is now an assistant professor in the Faculty of Education at Lakehead University. His research interests include relationships between environmental philosophy, ethics, education, and teaching.

ANN MATTHEWS is a doctoral student at the Ontario Institute for Studies in Education of the University of Toronto. Her doctoral work has focused on pedagogy and learning in higher education. She is currently working with Margrit Eichler (OISE/UT) researching adult learning in unpaid work.

HEATHER MENZIES is a writer, teacher, mother, gardener, and peace and social-justice activist. She is the author of eight books, including the 1996 bestseller *Whose Brave New World?* and, most recently, *No Time: Stress and the Crisis of Modern Life*. She dwells in Ottawa, Ontario, where she tries to live as much as possible on tidal time and tree time, not the clock and deadline time of cities.

ALEXANDRA (ALIX) MCGREGOR is assistant professor at the York University School of Nursing, Toronto, Ontario. Her teaching and research interests are focused on teaching-learning pedagogies, nursing ethics, and Heideggarian hermeneutic interpretive analyses of nursing students' narratives of their experiences in nursing.

LINDA MUZZIN is associate professor at the Ontario Institute for Studies in Education of the University of Toronto. She has written on the professions of medicine and pharmacy as well as the influence of the pharmaceutical industry on the university.

NATASHA MYERS' current research explores the visual cultures of contemporary biology. A PhD student in the Science, Technology and Society Program at MIT, she investigates the production and circulation of images in laboratories and classrooms and the emergence of new biological imaginaries with innovations in computer-intensive microscopy, modelling, and simulation.

PEGGY TRIPP is a professor in the Departments of Biology and Women's Studies at Lakehead University in Thunder Bay. From a background in forest genetics research in the forestry faculty, she has redirected her scholarly pursuits and teaching to environmental ethics, feminist science critique, and ecofeminism.

NJOKI WANE, associate professor at the Ontario Institute for Studies in Education of the University of Toronto, specializes in African women and spirituality, Indigenous knowledges, Black feminisms, anti-racist teacher education, and anti-colonial thought. A recently published article is "Black Canadian feminist thought: tensions and possibilities," in *Canadian Women's Studies*.

BARBARA WATERFALL is Metis/Anishinaabe. Currently on faculty in Social Work at Wilfrid Laurier University, she was previously faculty in the Native Human Services Program (Social Work) at Laurentian University. Finishing her doctoral thesis on the topic of Decolonizing Native Social Work Education, her research interests are in feminist anti-colonial thought, Indigenous knowledges, spiritual ontology, and multi-centric social work practice.

CREATIVE ARTISTS

EILEEN ANTONE is associate professor in the Department of Adult Education and Counselling Psychology and the Transitional Year Program of the University of Toronto. The primary focus of her work is with Aboriginal students in university studies. A member of the Oneida of the Thames First Nation, she has many years of experience with Aboriginal communities and organizations, advocating for Aboriginal perspectives.

Now living on the west coast, DIANE RAE is anchored in a life of art-making, having studied Fine Arts at the University of Toronto.

LORI FOX ROSSI is a professional nature photographer in Thunder Bay. She is best known for her assignment work with Canadian Geographic magazine. An ardent conservationist, she strives to be a voice for the preservation of clean air and wild places. Visit her website at www.lorifoxrossi.com.

KAREN SPINNEY-HELMER, formerly a biologist and instructor at Lakehead University, is currently pursuing her teaching and long-time passion for art from her home-based studio with a program called Art for Kids. Karen lives with her husband, two children, and numerous pets in rural Eastern Ontario.

LINDA STITT was born in Huntsville and raised in Georgetown in Toronto. In 1952 she moved to Thunder Bay, where she began the process of what Carl Sagan describes as "matter coming to consciousness." She has authored, or co-authored, ten books and is deeply concerned with the health of our little planet and all its inhabitants. Visit her website at http://www.geocities.com/passionate_intensity2002/

CHRISTINE VOGEL is an alumna of the Ontario College of Art in Toronto. A freelance graphic artist and teacher, Christine currently works as the Multimedia Design and Content Coordinator for Lakefield College School.

Prologue

I have chosen to interpret the term *Prologue* in its theatrical meaning – as opening reflections to be offered before the actual play begins.

I am very conscious of the generation gap between me and most of the authors of this volume. They are not just of the generation of my students, but some of them are my students' students. Thus, if this prologue is to be more than a ceremonial nod towards the past, it needs to concern itself with roots and with the soil in which the authors' works have grown.

Underlying all personal encounters and influences, there exists a common soil, a collective pool of feminist experience and achievement, that has nourished my young sisters to the point of their teaching in established institutions and being able to reflect on the content, method and thrust of their activities.

Today *equity* is not merely a dream or a social construct; it is now a legal concept, containing expectation and entitlement – this has not always been so. By the same token, *environmentalism* (i.e., "advocacy for or work towards protecting the natural environment,") has only relatively recently become a curriculum subject.

The seeds of these movements, though, have long been present in our common soil, more often than not sown by women.

As part of my prologue to this book, let me highlight two components of this soil: women's struggle for peace and, contained in it, their struggle against nuclear war, nuclear weapons, and nuclear energy.

The long and creative work of women for peace and against war and violence is part of our common feminist experience. It is work that reminds us how women have transcended boundaries of nationalism, class, and "race" in their advocacy for a just and peaceful world. Their understanding of how an equitable world could be structured is well documented (Bussey et al. 1965).

In Canada, women's struggle for equity and women's activism for peace have been intimately linked (Kerans 1996). Within these developments, the anti-nuclear movement has provided a particularly important practicum of feminist teaching and activism.

In the 1960s, organizations such as Voice of Women made submissions to Parliamentary committees on systemic barriers to equity for women at the same time as they researched and presented briefs on radioactive fallout monitoring, on the health effects of Sr 90 or radioactive Iodine on children (all children, all over the world), thus making ignored or uncomfortable research findings publicly available (Macpherson 1994).

Discussions of this new evidence – such as those in Rosalie Bertell's documentation of the effects of low level ionizing radiation (Bertell 1985) or the Helen Caldicott film "If You Love this Planet" (Caldicott 1978) – in innumerable church basements and classrooms brought this new scientific knowledge and political activism to Canadian women.

They, in turn, became teachers in their communities, teachers who understood the essential links among all living beings, and made this insight the centre of their personal and political understanding as well as part of the common soil.

Today, it is interesting to recall the debates on the safety, cost, and reliability of nuclear power that took place during the past four decades, because the argumentation is so similar to the current discourse on genetically modified seeds or food.

When scientists like Dr Bertell or I argued the importance of long-term effects of radiation, the intrinsic difficulties of nuclear waste storage, and the miscalculation of the costs of nuclear power, there was usually a standard set of responses: What we presented was either "shoddy science," "an unrepresentative situation," or "a lack of confidence in future scientific problem solving." And, of course, there was the ever-present rhetorical question: "Do you want to deprive the Third World of this cheap and reliable source of energy?"

Forty years later, nuclear reactors are expensive and dangerous burdens, while the same arguments of shoddy science and disregard for the needs of the hungry are levelled against those who question the promotion of genetically modified plants.

To me, this correspondence indicates that present teaching and activism could benefit from remembering and re-evaluating past feminist teachings in the public sphere. It may not be particularly cheering to realize how often feminists have been right. They were not only right on social and political grounds, but they were also correct in terms of facts and predictions. Yet it may be necessary to stress such evidence, because the changes needed for the world to survive, are *systemic* changes.

The specific issues addressed in our teaching and activism have to be seen as illustrations of more basic problems. These problems can be resolved only through a fundamentally different ordering of social and political powers and priorities. The essentially feminist ordering required is built on the cardinal tenet of equality, that *all in the biosphere are entitled to equal care and concern*.

It is the practical execution of this principle that will ensure respect, survival, and peace. The wide spectrum of women's attempts to develop appropriate practices and to decline participation in inappropriate and destructive ones constitutes the foundation of feminist teaching and activism — past, present and, likely, future.

This is where our roots are and the source of our strength is found.

Ursula M. Franklin OC, FRSC
October 2003

REFERENCES

Bertell, Rosalie, 1985. *No immediate danger.* Toronto: Women's Educational Press.

Bussey, Gertrude and Margaret Tims, 1965. *Women's international league for peace and freedom 1916–1965, a record of fifty years' work.* London: Allen and Unwin.

Caldicott, Helen, MD, 1978. *Nuclear madness, what you can do.* New York: W.W. Norton. Also see Caldicott 1992. *If you love this planet.*

Kerans, Marion Douglas, 1996. *Muriel Duckworth, a very active pacifist.* Halifax: Fernwood Publishing.

Macpherson, Kay, 1994. *When in doubt, do both.* Toronto: University of Toronto Press.

TEACHING AS ACTIVISM

Overview

PEGGY TRIPP AND LINDA MUZZIN

SPRING AND HOPE

We have had trouble deciding on an appropriate title for this book. We originally considered naming it *Teaching (Science) As If the World Mattered*. The title reflected the fact that we were working in bioscience faculties where reductionist views about how science should be pursued were dominant. Each of us taught and did research in ways that challenged that reductionistic mould. But, since the term *science* has all the same semantic baggage as the positions we critique, and since any knowledge that we would advocate would be holistic rather than reductionistic (reductionism involving the exclusion of our social world and its worldly significance), we have dropped the word *science*. The working title during the writing of the book was *Teaching As If the World Mattered* with a subtitle, *Equity Meets Environmentalism*.

As a group of equity theorists, scientists, and environmentalists, we believe that there is an important parallel between activisms for the environment and for equity and so we finally decided on *Teaching as Activism* as the title. What is teaching as activism? We believe that our teaching is part and parcel of the activism expressed in the anti-globalization demonstrations such as those in Seattle and Quebec City. It includes the teaching that we professors do inside and outside the classroom. This book promotes teaching about knowledges that have been marginalized in academia. These include spiritual and other non-rational knowledges as well as many feminist ideas about science and the broad field of anti-colonial thought. The two major currents of activism, equity and environmentalism, permeate our lives, our work, and our thought. Thus they are prominent in this collection.

Feminist environmentalists have long recognized that speciesism, or discrimination against non-human species, is an important part of planetary degradation and this position is recognized by many authors in this collection. In a parallel movement, feminists and anti-racists have argued that inequity, or discrimination against particular groups of humans, is the basis of war and poverty. Highlighting this parallel is an intention of this collection. In the past few years, writers about Indigenous knowledges have pointed to crucial intersections in these arguments. For example, the spirituality of Aboriginal peoples has been based on respect for the earth; and thus their subjugation goes hand in hand with disrespect for life. Although critiqued for their essentialism, ecofeminists have similarly argued about women's respect for the earth. Even though individual authors, embedded within their intellectual disciplines, do not all integrate these two themes in their writing, the signposts pointing towards these intersections are conspicuous in our collection. Thus, this book provides a forum in which these 'two solitudes' of scholarship can begin to speak to one another. Although as a group, we have been meeting face to face and talking to one another for years, we still fall short of integration of these parallel movements in our work. Thus, this volume needs to be seen as a work in progress towards the rapprochement we yearn for between environmentalism and equity studies.

Theoretically and methodologically, the pathway into this rapprochement comes from our personal experiences. As editors, we instructed authors to begin with narratives locating their own consciousness about activisms for equity or environmentalism in their life work. In our own narratives, for example, we realized that gardening is a shared avocation. Naturalistic gardening is a metaphor for appreciating diversity and showing respect for life. It can be seen as a kind of activism as can be many practices not often considered as activist. In the case of gardening, nurturing our appreciation of diversity takes a very concrete form in Spring, that is, working with the soil. With gardening, we are able to appreciate our privilege in having access to land, our closeness to nature, our learning about native plants, and our recognition of the spiritual component of our work. From my perspective (PT), it is spring and I am busy planning new additions to my boreal garden in Thunder Bay, Ontario. These are problematic, in that I am committed to not plundering them from the indigenous locations to which they still cling. They survive despite the devastation that came as part of the White habitation of Canada and the readily apparent changes associated with global warming. The regrowth each spring of the native bunchberry, lady's slipper, and bluebead lily provide me with hope for northern forest ecosystems that are being altered on such a vast scale. Our penchant for paper, such as this book in your hands, is creating a transition in our forests from those dominated by softwoods such as pine and spruce to a simpler system dominated by the hardwood, aspen.

As for me (LM), I have made a commitment to encouraging indigenous plants in my southern Ontario home. This involves ordering seeds, the names of which I gleaned from Peggy's catalogue of native plants. My project this summer, just as Peggy's has been for most of her life, is to begin to do something personal to address what Wood (2000) has called "agricultural colonization and landscape creation." This was the process by which settlers in the nineteenth century destroyed southern Canada's Indigenous plants, replacing them with Euroweeds that have overrun the native plants, killing off the food of native birds and hosts of Native butterflies and moths. Fortunately, the bright yellow sheaths of the skunk cabbage, the harbinger of spring, still poke their enormous leaves through the black wetlands at the back of my property. The wild white trilliums, milkweed, and purple coneflowers that grew there a decade ago have disappeared, but wild strawberries, bright red bee balm, black-eyed Susan, purple bottle gentian, numerous ferns, yellow evening primrose, Solomon's seal, daylilies, goldenrod, mint, bleeding heart, wild geranium, and wood sorrel have survived. Over all of these tower four great indigenous white pine trees that are approximately 150 years old. There is still time and still hope.

LOCATING OURSELVES AND OUR WRITING

Most of the authors who contributed to this book are part of a research group that has been meeting regularly since 1998 both at scholarly conferences which we have organized, as well as in a smaller group. We have thus become familiar with one another's thinking and missions and the ways in which they are interrelated. This means that, in most cases, the contributions here are in keeping with the positions that members of our group have repeatedly articulated to us and which we encouraged them to crystalize succinctly for this volume. Their chapters, if you will, are their lifeworks and the projects to which they have been devoted during the time that we have known them. Thus, although the messages are all different about how and what we should be doing in our activist teaching, we believe that they are all of one cloth from the senior to the junior.

We have told the story of how two White feminists came to address these matters elsewhere (Muzzin *et al.* 2002; Muzzin *et al.*1997). Basically, we founded a network of about 100 professors of science and social science and their graduate students called *Biology As If the World Mattered*. From the beginning, the group called itself by the acronym BAITWorM. It was fortuitous synchronicity that Ursula Franklin, who was the recipient of our first "Worm of the Year Award," has long been associated with the metaphor of the earthworm in her peace activism. Peggy Tripp and Ursula Franklin had met during the filming of *Asking Different Questions: Women and Science* (1996), which focused on their lifework along with that of two other feminist scientists, Rosalind Cairncross and Karen Messing. Karen Messing and her friend Margrit Eichler later joined BAITWorM, and we formed a core of researchers and teachers in science, social science, and women's studies, who functioned as a core group. *Asking Different Questions* had several messages that we have carried forward in the network: we need to return to first principles in constructing knowledge by focusing on the spirituality of nature; we need to problem-solve using community approaches within indigenous localities; and we need to resist the destructive force of corporate capitalism in its attempt to control scientific praxis. This book grew out of the first two conferences which we hosted at OISE/UT in 2000 and 2001, but it is not a set of conference proceedings. This would have been impossible, in any case, since 140 papers were presented.

After our first conference, we realized that we were dominated by White feminists and environmentalists. Thus our second conference was much more diverse. The authors of this volume represent that diversity. For example, we are of many 'races.' This word appears in single quotation marks to emphasize that it is a social construction, having no biological basis. However, it is real, and is a major component of our social world. As such, it needs to be named. Those of us who are White belong to the 'race' that has built a hegemonic androcentric scientific knowledge that threatens to destroy the earth. We are all complicit, with no one escaping that problematic. However, we stand together in our commitment to the tearing down, through our teaching and knowledge construction, of oppressive practices that promote a monocultural White knowledge at the expense of Indigenous knowledges and respect for many ways of knowing. As one of our Aboriginal colleagues is fond of pointing out, we all had Indigenous origins at some point. A common weave is our feminism, which is of many strands. We are women and men, young and old, student and teacher, all speaking out for the enactment and preservation of the diversity of our human ways, as well as the ways of the other living Beings with whom we share the earth. This last is not said lightly, since we are painfully aware of the scientific and social scientific hegemony that would make invisible the circumstances of our being on this earth and would crush our respect for, and attempts to co-exist with, other species.

Although we believe that this book is unique, we must acknowledge writing that has come before. We firmly believe that one's pathway to an intellectual insight is related to one's original position. In writing this overview we have come to realize that we inhabit different universes in naming what has been influential in our own thought and the thought of those making this collection.

In Peggy Tripp's experience, the single book that opened the door to her critical thinking about teaching science in general and feminist science critique in particular, was Evelyn Fox Keller's classic book *Reflections on Gender and Science* (1985). It remained on her bedside stack of books for months in spite of many aborted attempts to grasp the main arguments. When its messages finally penetrated her psyche, Peggy was startled by the sense of hearing another voice expressing her own deep-seated sentiments. From that point, she read the feminist science genre avidly. Numerous works appeared as collections in the late 1980s and early 1990s, containing writing by white feminists such as Dorothy Smith, Emily Martin, Sandra Harding, Donna Haraway, Ruth Hubbard, and Elizabeth Fee (e.g., see Harding and O'Barr 1987). Recent publications of readers in the area of gender and science indicate the continued popularity of this genre, both in the republication of older classics along with newer contributions (e.g., Keller and Longino 1996; Lederman and Bartsch 2001).

Since Linda Muzzin has been teaching sociological and educational theory, her "classics" include many of the sociological 'great books' as well as more recent critiques of globalization (Amin 1997; Teeple 2000). Like a generation of sociologists in Canada, she was inspired by Dorothy Smith's writing, which was an excellent introduction to critical feminism (e.g., 1987, 1998), and Sandra Harding's consideration of gender, racial biases, and postcolonialism in science (1993, 1998). Also important is the much-critiqued, but classic, feminist epistemological work, *Women's Ways of Knowing* (Belenky *et al.*, 1986) which was the first pathway into learning about difference for some of us White feminist teachers, along with Sue Middleton's *Educating Feminists: Life Histories and Pedagogy* (1993). It was Middleton's reproduction of her childhood drawings made growing up in New Zealand, of the ships of the Euroexplorers who 'took possession' of previously inhabited territories that awakened Linda to our complicity in the colonial project of subjugating Indigenous knowledges. Finally, a few excellent collections focusing on equity in the academy have appeared in the past few years (e.g., Roman and Eyre 1997, Richer and Weir 1995).

But it was BAITWorM discussion that changed our focus towards the anti-racist and anti-colonial writings that have taken centre stage in equity studies in recent years. Vandana Shiva's early writings introduced the term "maldevelopment" to underscore the class oppression that accompanied the development projects in India (1989, 1993, 1997). She is an intellectual leader for her work at the intersections of sexism, classism, and racism. Other books that deal with these intersections are Bannerji's *Thinking Through* (1995) containing essays on feminism, Marxism, and anti-racism and the recent Oxford series collection of contemporary writings entitled *Feminism and 'Race'* (Bhavnani 2001).

It is one thing to be awakened to the possibility of a feminist science; it is quite another to act on this realization. In the realm of equity, Mahatma Gandhi (2002) and Martin Luther King modelled passive resistance. In the realm of resisting environmental degradation, Lois Gibbs (1998) modelled persuasion and persistence. Linking White corporate capitalist monoscience to neocolonial oppression and the dominance of reductionist ways of knowing, Vandana Shiva, Bina Agarwal, (2001) and Esther Wangari *et al.* (1996) connect equity with environmentalism and invite us to pursue ways to dismantle this neocolonialism in our teaching and research.

It was also the BAITWorM group that made us realize that equity and anti-speciesism arguments were parallel. Ecofeminist literature is the one source that brings these ideas together, at least with regards to sexism. The two early collections by Plant (1989) and Diamond and Orenstein (1990) are acknowledged as inaugurating the emergence of ecofeminism as an academic field of study. More recent publications by Carolyn Merchant (1995), Chris Cuomo (1998), and Val Plumwood (2002) have added theoretical depth to the field. Except for Shiva's work, this literature needs to be racialized.

Beyond feminist writing, the focus on speciesism is prevalent among writers who might be recognized as 'environmentalists.' First and foremost is Rachel Carson, whose book *Silent Spring* (1962) introduced North

America to the significance of our environmental interconnected web. The deep ecology movement inspired by Arne Naess (1989) further developed this environmental ethic away from human-centred values towards biocentrism. These foundational writings opened the flood-gate to environmentalist literature far too extensive to list. What they have in common is an anti-speciesist message, that is, a plea for an attitudinal paradigm shift to appreciate species other than our own as morally significant.

As befits a collection about teaching as activism, the papers in this book are also inspired by the pedagogical literature (see Ann Matthew's reference list in this collection for a selection of the numerous books about engaged and transformational feminist pedagogy). There is also Paolo Freire's work including the reprinted *Pedagogy of the Oppressed* (1995) and bell hooks' *Teaching to Transgress* (1994). Pedagogical writing about anti-speciesism converges in the classic works by David Orr (1992, 1994) who writes passionately about ecological literacy in the education system. Canadian contributions to the field include the *Canadian Journal of Environmental Education*, edited by one of the contributors to this volume, Bob Jickling.

Finally, books that feed the spirit are crucial. It is not surprising, since we are academics, that such books assisted in our reawakening, and that our way of expression is in book form. Of the books that reawakened Peggy, most notable is Simon Schama's (1996) *Landscape and Memory*. For Linda, the books of Courtney Milne, the Saskatchewan photographer, have been continuously inspirational. In particular, the enchanting *Sacred Places* (1991) is always in her hands, her heart, and her thoughts. In fact, our experiences with such inspirational books led us to understand how the integration of text, photographs, drawings, and poetry are integral to our reawakening to the possibilities of the world as we walk among the rocks, trees, and waterfalls. Thus we have encouraged the contributors to this book to also include non-textual representations.

ORGANIZATION OF THE BOOK

This book is divided into three parts: Personal Experiences and Paradigm Shifts, Problematization of Dominant Realities, and Weaving New Worlds and Reclaiming Subjugated Knowledges. We see these divisions as representing the significant events in the process of teaching and doing research as activism. These three processes were identified in our instructions to authors. We requested that authors describe becoming aware of problems in their own experience (Personal Experiences and Paradigm Shifts), reading and critiquing (Problematization of Dominant Realities), and applying their insights to their practice of teaching and research (Weaving New Worlds and Reclaiming Subjugated Knowledges). The reference to "Reclaiming" in the third section means taking back what is yours. This process is basic to feminist and anti-racist writing and involves decolonizing knowledges once consciousness has been raised. Several of the writers in the last section attempt to weave new worlds in the academy, the professions, the laboratory and the community by applying their insights to knowledge construction.

Personal Experiences and Paradigm Shifts

How is consciousness "raised" about environmental, ethical, and social issues? As feminists, we have learned that the personal is political, and so we know that there is a personal and spiritual component to consciousness raising. Let us give an example from the life of our group. Meeting on our first of several occasions in concrete buildings in Toronto severed the group's crucial connection to nature. Thus it was with great joy that we invited our colleagues to join us two summers ago at Ouimet Canyon, an immense rocky chasm formed by a rift, which winds its way over to Lake Superior, the largest fresh-water lake in the world. Greeted even in the heat of summer by an upward breeze of refrigerated air, since the bottom of the canyon is so deep, the young and old among us scampered among the rocks and trees there, reconnecting with the spirituality that places such as Ouimet bring. That afternoon, sitting beside a lake on the Sibley peninsula, talk turned to spirituality. Later that night, relaxing at Peggy's home, we became deeply engaged in a debate about whether the 'new' genomics really was new or old. We felt alive, engaged, and on the edge of transformation. It is this spirit that we hope is present in this book.

The first section contains papers that express a yearning for the world to be different. They describe various kinds of awakening, but most particularly awakening to the limitations of an anthropocentric world in which humans are valued over animals, trees, and rocks. The authors of the papers in this section are all concerned with environmental ethics, with wolves and trees and the forest. According to Leesa Fawcett who discusses her encounter with a furry bee/hummingbird/moth she met in her garden: "The mystery of meeting an other expanded my world. Filled with awe at the sphinx moth's proximity to my daily life, with the knowledge that our lives intersected, even if only momentarily, there was a symmetry between our worlds. To observe, to give attention to another life is no small task in these days of hectic, frantic activity. It is in the fullness of such attention that possible new ethical relationships lie" (2000, 135–6). Fawcett's point is that in imagining a world that matters, there is sanctity in each blade of grass (Campbell 1973).

In this section, Heather Menzies' paper uses the metaphor of a field of tree saplings to represent her awakening to a world in which both humans and other life forms matter, and where the best way to know the world is to engage in a respectful dialogue. Heather is a prolific writer who has been inspiring us to awareness of our technological and cultural baggage. In her paper, "The Seedlings Mattered," she describes a pivotal experience in her youth of witnessing the failure of seedlings to thrive. Her activism is expressed in her argument of the necessity to participate in relationships in order to learn – a message she learned from an Aboriginal Elder. Bob Jickling's chance encounter with a wolf on a canoe trip provides the backdrop for his troubling paper about the canicide or premeditated slaughter of wolves that is a result of our 'wildlife management' strategies of the North. The ponderings of Bob Jickling (as well as Peggy Tripp in this section) emphasize that our environmental ethics, as the foundation of our scientific knowledge construction, need re-examination. That is, environmental ethics need to be foregrounded for students. His activism is expressed by challenging us to practise ethics as an everyday activity and to reincorporate ethics into our science teaching. Like Menzies', his chapter deals with a non-human-centred paradigm as well as with feelings and emotive reasoning. As such it leads into the next paper by Wane and Waterfall that focuses on spirituality. This chapter begins to explore the complexities of Indigenous knowledges by bringing together the writing of two Indigenous traditions from different parts of the world – eastern Africa and northeastern America. Njoki Wane and Barbara Waterfall are gifted teachers and writers, awakening students at OISE/UT and Laurentian University respectively to the inherent spirituality of Aboriginal thought and practice. In their paper, "The Hoops of Spirituality," they explore each other's traditions, and focus on how Aboriginal spirituality that inherently recognizes the sanctity of all life can be brought into the classroom. Making spirituality central in their teaching is activism. The fourth paper, by Peggy Tripp, relates her professional challenge of closing down her traditional science laboratory, an act of resistance. Instead of teaching about 'improving' trees, she returned to long-held notions of sustainability in her teaching within a traditional science and professional curriculum. In this paper we follow Peggy Tripp's process of "awakening" to the ethical dimensions of her discipline – dimensions that had been ignored in order to succeed in academia. Further, she describes her vision of what it means to her to teach sustainability. Like Wane and Waterfall, Tripp introduces spirituality into the classroom.

We pointed out at the beginning of this overview that one purpose of this book is to explore the intersections of speciesism and other difference. Taken together, the first four papers in this section address this too-often neglected topic of specieism as well as illustrating the spiritual nature of consciousness-raising.

Moving from environmental to equity teaching, the gendered nature of professional education and practice is unpacked in a paper by Moira Grant, who traces the history of subjugated medical laboratory technicians. Moira, a medical laboratory technologist herself, also describes her awakening to feminist and other critical literature, an awakening which allowed her to problematize the position of women in her profession. In the sixth paper, Ann Matthews draws our attention to the fact that feminists have developed a substantial body of literature describing their unique teaching and activism philosophies and practices that can inform teaching in higher education (e.g., Brookes 1992; hooks 1994; Lather 1991). Reflecting on her experience as a community college teacher, Ann describes her discovery of transformative (feminist) teaching. Incorporating these practices into her teaching was a radical act for her, as it would be for anyone. The chapter emphasizes the broader definition of pedagogy as "the nature of knowledge [including how] knowledge is produced, negotiated, transformed, and realized in the interaction between the teacher, the learner and the knowledge itself" (Kenway and Modra 1992, 140). Imagine how mainstream teaching would be changed if everyone employed these pedagogies.

As the final paper in this section, Marianne Ainley's work genders and begins to racialize her previous accounts of the role of women in knowledge production in Canada (1990). In the present collection, using an autoethnographical methodological approach, Marianne Ainley reveals both her awareness of sexism and racism within existing scholarship on science and illustrates how she reclaimed intellectual territory that she had lost in her teaching and research. This includes women's relationships with their environments and the work of women scientists. Her paper suggests the intersections of gender and race and the construction of knowledge, including the unacknowledged appropriation of Aboriginal knowledge into Western science.

Problematization of Dominant Realities

While the papers in the first section dealt the with paradigm and consciousness shifts that precede activism teaching, this section surveys the relations of ruling that permeate teaching. The four papers in the second section all deal with discourses and structures, including capitalism, that saturate our scientific way of knowing. This economic reality needs to be named and problematized in teaching and knowledge construction in higher education as if the world matters. The word "problematizing" refers to critiquing dominant discourses; it was popularized by Dorothy Smith in her writing (1987, 1990). In this text the problematization of a term is indicated by single quotation marks. In the tradition started by Marx and carried on by feminists and anti-racists, the process of problematizing makes visible what Dorothy Smith calls 'relations of ruling' that are part and parcel of the powers that attempt to control the construction of knowledge. What are the 'relations of ruling?' Recent writing on the globalization of capital maps a thrust towards the dismantling of the welfare state and towards the enshrinement of corporate rights in binding global legislation (Teeple 2000; Barlow and Clarke 2001). Various strategies for addressing these initiatives have appeared (e.g., Amin 1997; Klein 2000). The four selections in this section all problematize the globalization of corporate economics, with the first critiquing Eurocentrism, the second globalization, the third higher education, and the fourth science.

Once aware of discriminatory practices in our thinking and knowledge construction, where do we turn for guidance? In this second section, we emphasize that it is not necessary to re-invent the proverbial wheel. Inspirational ways of knowing are close at hand. Post-colonial and anti-colonial writing in the last decade has made it clear that many alternatives to White monocultural thought have always been readily available (T. Smith 1999; Dei et al. 2000). White Western writers in higher education have been agents of subjugating these knowledges. This subjugation, in which we are all complicit, must stop. Feminist writers and writers on Indigenous knowledges encourage, for example, women and Indigenous peoples to reclaim knowledges that have been made invisible in mainstream thought in higher education.

Anti-colonial and post-colonial activism involves both 'tearing down' and 'building up.' The four papers in this section all emphasize this point. Marie Battiste's "You Can't Be the Global Doctor if You're the Colonial Disease," begins with a powerful indictment of Eurocentric domination of knowledge production. Her paper places knowledges of the land at the centre, where they belong. The paper is a concise statement of her inspirational writing encouraging the protection of Indigenous knowledges from their appropriation and commodification by dominant discourses (Battiste and Henderson 2000; Battiste 2000). At the same time, we catch a glimpse of how the spirituality of people living close to the land exemplifies their respect for other life. In Battiste's words: "It is not helpful to attempt to define Indigenous knowledge, for it is a comparative knowledge system that should not be pounded into Eurocentric categories. Indigenous knowledge involves ecosystems and Indigenous peoples' relationships with them. Indigenous knowledge allows those who move within it to discover everything about the world they inhabit that they can discover with the tools of language and culture."

Proceeding from the post-colonial argument presented by Battiste, Vanaja Dhruvarajan's paper critiques both British colonialism in India and the neo-colonialism of globalization discourses. That is, she critiques the contemporary form of capitalism – globalization – as a new form of colonization. She explores constructions challenging the neoliberal paradigm, including the work of Walden Bello, who has said, "We are talking ... about a strategy that consciously subordinates the logic of the market, the pursuit of cost efficiency to the values of security, equity and social solidarity. We are speaking, in short, about re-embedding economy in society, rather that having society driven by economy" (2000, 11). Throughout her career, Vanaja has written about women of colour and anti-racism, a focus of interest that has led to her critical examination of globalization appearing here.

The remaining two papers in this section focus on *the* knowledge that is being promoted in the process of globalization – biotechnology (or engineering life). In her paper, Linda Muzzin argues that the pharmaceutical sciences are at the centre of academic capitalism and she urges academics to acknowledge that higher education is not just complicit in speciesism, sexism, and racism, but a major agent in the perpetuation of inequity. The main point of her critique is that the globalization of education involves the marketing of biomedical knowledge systems to the rest of the world.

In "Working in the Field of Biotechnology," Elisabeth Abergel describes her discomfort as a scientist working in the 'new world order.' As a recently trained scientist turned faculty member, she links the degradation of environmental systems to the failure of democratic institutions to provide the rationale for a biotechnologically defined sustainable future.

This chapter follows on the heels of her doctoral thesis on the 'canola story,' constructed by the Canadian government and pharmaceutical industry alike as a globalization 'success story,' while it should be seen as a warning about the dangers of commercial tampering with life.

Weaving New Worlds and Reclaiming Subjugated Knowledges

Given the growing awareness of power structures within higher education that support various 'isms' – from speciesism to capitalism – how can we trope them to teach and do research as activism for change? Not every strategy that we employ will be successful on the first try. Further, the insidious complicity with the status quo that characterizes our work in higher education can easily undermine our initial attempts at reorientation. Clearly, the ideas that need to be put forward for addressing these issues need to be big and they need to be various.

The first three papers in this section all propose big ideas – from Margrit Eichler's call for the re-education of social scientists to Alexandra McGregor's proposal for revision of the whole system of evaluation that we use. Margrit Eichler argues for a basic change in the content of the social sciences as a group of disciplines. She challenges scholars to teach the relevance of sustainability in social science, and suggests ways in which social scientists can optimize their literacy with respect to environmental sustainability. This insight by Margrit Eichler, a foremost scholar engaging with sexism and racism (Eichler 1991, 1997, 1998), has emerged with her realization, central to this book, that sustanablility and social justice need to be connected with one another. The most androcentric of professions (engineering) is taken on by Monique Frize, appointed one of the five first NSERC chairs in women in science and engineering in Canada. In her upbeat chapter, she suggests that a necessary first step towards addressing the problems within this discipline is to improve the climate for women and other underrepresented groups. We would add that substantial reconstruction of this field is necessary to address sustainability issues such as the effect of engineering projects on life. Indeed, Frize brought together hundreds of women engineers and scientists in the Fourteenth International Conference of Women in Engineering and Science in July of 2002 to share their work on solutions for a variety of major environmental problems.

Alix McGregor's paper points to the contradiction between student-centred pedagogy and traditional evaluation systems. This chapter goes to the heart of what makes our educational and knowledge production system authoritarian by suggesting that a caring pedagogy would have a caring evaluation system. McGregor's doctoral thesis was a careful study of the management of failure in nursing which has led to her current interest in evaluation in higher education. As an award-winning teacher, McGregor reflects on her personal experiences and tells a tortured tale of engaging with an uncaring evaluation system.

Are there any inspirational examples of teaching as activism? The last four papers in the collection present full blown visions of transformative teaching and research in that they identify processes that have been demonstrated to *work*. The word *transformative* has a long history in the literature on education (e.g., Shor and Friere 1987; Simon 1992). Teaching and doing research, the activities that are at the core of our being in higher education, are also at the core of our praxis, and we assume, the praxes of those who will read this collection.

In the last few years, we have witnessed an exponential increase in writing about Indigenous knowledges that has made it one on the most exciting 'new' topics in the humanities and social sciences. Concluding her argument made earlier in this collection, Marie Battiste calls for approaches to globalization that will encourage intercommunication among Indigenous peoples and resistance of the commodification that globalization imposes on them. She mentions the success of Maori efforts in the reclamation of subjugated knowledges by this Indigenous population. Robin Cavanagh's paper proposes an Aboriginal Education Framework, a transformative pedagogy based on the teachings of Elders. This Framework, the result of his personal transformation, draws the teacher and learner together in a reclaimed Aboriginal knowledge system where a teaching wand is used to show students how to live as if the world mattered. Next, Natasha Myers' paper, entitled "Visions for Embodiment in Technoscience," addresses what may be the most difficult nexus of all, the scientific laboratory. Displaying all the dexterity of the dancer she is, Myers brings us back to the consideration of speciesism that began this collection. She extends Merleau-Ponty's concept of the flesh and Donna Haraway's ideas about situated knowledges in her proposal for a new scientific subjectivity that would provide a caring or embodied relationship with the object of scientific scrutiny. Thus she advocates bringing spirituality into the laboratory by the use of "insurgent" technologies. And finally, the method and substance of Leesa Fawcett's teaching about bioregionalism demonstrates the engagement of students in their natural surroundings of the Niagara escarpment. Consistent with our interpretation of teaching as activism in this collection, Fawcett refers to her teaching as "subversive science."

BUILDING BRIDGES:
THE CHALLENGES OF INTERDISCIPLINARITY

We have thought long and hard about why, although the equity theorists in this collection are clearly making parallel and interconnecting arguments with the environmentalists, very few single authors cross this divide to discuss both equity and environment. The answer lies in academe's tradition of dividing knowledge: if one is a scientist (Tripp, Fawcett, Jickling, Abergel) or a science-based professional (Grant, Frize, McGregor), the nature of science education and expertise necessarily puts one at a disadvantage for being an expert also in the humanities and social sciences. The scientifically oriented authors we have included rely on feminism, ethics, and to a lesser extent, anti-colonialism to make a connection. From the other side of the divide, those with social science expertise (Eichler, Muzzin, Menzies, Dhruvarajan, Ainley) have difficulty being scientists as well; these authors bridge this divide through feminist science critique, historical methods, personal experience that contradicts dominant paradigms, and to a lesser extent, anti-colonial ideas. Indigenous knowledge scholars are located at a place where the "social" and the "scientific" are not divided and so their contributions are models for bridging this divide.

WHY ANOTHER BOOK?

To summarize, our answer to why another book in view of all the writing about engaged pedagogy as well as all the anti-racist, feminist, anti-globalization, and anti-colonial writing, is that unfortunately, in our view, few collections bring all of these areas together. We think that the pedagogical literature is a subjugated one, as far as feminist critique of science is concerned. This is particularly problematic since feminist pedagogy also entails knowledge construction. Further, we want to introduce an emphasis on higher education to the field.

There are at least four ways in which this book differs from other books in print. First, we believe that the intersection between engaged pedagogy and what we have come to see as a post-colonial vision of science needs to be explored. Further, much of feminist pedagogy is White. It thus does not engage with the critique of science that is so prominent in the growing literature on Indigenous knowledges. A second and related way in which this book differs from other books is that it brings together Aboriginal and White writers, each of whom are working in their unique settings teaching a message of hope and a new beginning. Thus our hope is that this collection will engage readers interested in environmental justice and feminism who might not otherwise imagine the global context of their teaching and research. Third, the consideration of anti-speciesism is conspicuous for its absence in much of the anti-globalization, anti-colonial writing. We are committed to focusing awareness on the dual oppression of human groups as well as non-humans, including the environment.

The final way in which this book differs from others is in spanning the range of interdisciplinary feminist and anti-colonial scholarship in which we are actively engaged. This brings, we believe, an immediacy to our writing that foregrounds the specifics of each area. This engagement in the local situation of knowledge construction across diverse areas is critical. For Peggy Tripp, for example, the concern is with forestry science. For Elisabeth Abergel, it is genomics, and for Linda Muzzin, the pharmaceutical industry. For Monique Frize, it is engineering; for Alix McGregor and Moira Grant, the nursing and medical laboratory technology professions respectively. For Vanaja Dhruvarajan, the focus is on India. For Njoki Wane, it is eastern Africa. And for Marie Battiste, it is the Indigenous knowledges of the region called Canada. And for virtually all of us, our concerns are centred in the classroom, though we also realize the potential for longer-term changes in graduate and undergraduate curricula, programs, and evaluation procedures. We also recognize the potential for broader changes when we consider life-long learning for adults, for teachers of primary and secondary students and the construction of knowledge for future generations.

REFERENCES

Agarwal, B. 2001. Environmental management, equity and ecofeminism: debating India's experience. In *Feminism and 'race' (Oxford Readings in Feminism)*, ed. K-K. Bhavnani, 410–455. New York: Oxford University Press.
Ainley, M. ed. 1990. *Despite the odds*. Montreal, PQ: Vehicule Press
Amin, S. 1997. *Capitalism in the age of globalization*. London: Zed Books.
Asking different questions: Women and science. 1996. Directed by Gwynne Basen

and Erna Buffie. 57 min. Artemis Films/National Film Board of Canada.. Videocassette.

Bannerji, H. 1995. *Thinking through: Essays on feminism, Marxism and anti-racism*. Toronto: Women's Press.

Barlow, M. and T. Clarke. 2001. *Global showdown*. Toronto: Stoddard.

Battiste, M. ed. 2000. *Reclaiming Indigenous voice and vision*. Vancouver, BC: University of British Columbia Press.

– and J.Y. Henderson. 2000. *Protecting Indigenous knowledge and heritage: A global challenge*. Saskatoon, SK: Purich Publishing.

Belenky, M.F., B.M. Clinchy, N.R. Goldberger, and J.M. Tarule. 1986. *Women's ways of knowing*. New York: Basic Books.

Bello, W. 2000. *From Melbourne to Prague: the struggle for a deglobalized world*. Talk delivered at a series of engagements on the occasion of demonstrations against the World Economic Forum in Melbourne, Australia, Sept. 6–10.

Bhavnani, K-K. 2001. *Feminism and 'race' (Oxford Readings in Feminism)*. New York: Oxford University Press.

Brookes, A-L. 1992. *Feminist pedagogy*. Halifax: Fernwood Publishing.

Campbell, M. 1973. *Halfbreed*. Lincoln NB: University of Nebraska Press.

Carson, R. 1962. *Silent spring*. Boston: Houghton Mifflin.

Cuomo, C. 1998. *Feminism and ecological communities*. New York: Routledge.

Dei, G.J.S., B. Hall and G. Rosenberg, eds. 2000. *Indigenous knowledges in global context*. Toronto: University of Toronto Press.

Diamond, I. and G. Orenstein, eds.1990. *Reweaving the world*. San Francisco: Sierra Club Books

Eichler, M. 1991. *Nonsexist research methods. A practical guide*. New York: Routledge.

– 1997. Feminist methodology. *Current Sociology* 45(2):9–36.

– 1998. Towards a more inclusive sociology. *Current Sociology* 46(2):5–28.

Fawcett, L. 2000. Ethical imagining: Ecofeminist possibilities and environmental learning. *Canadian Journal of Environmental Education* 5:134–149.

Freire, P. 1995. *Pedagogy of the oppressed*. New York: Continuum.

Gandhi, M. 2002. *The essential Gandhi*. New York: Vintage Books.

Gibbs, L. 1998. *Love Canal: The story continues ...* Gabriola Island: New Society Publishers.

Harding, S., ed. 1993. *The "racial" economy of science*. Boomington: Indiana University Press.

– 1998. *Is science multicultural?* Bloomington: Indiana University Press.

– and J. O'Barr, eds. 1987. *Sex and scientific inquiry*. Chicago: The University of Chicago Press.

hooks, bell. 1994. *Teaching to transgress*. New York: Routledge.
Keller, E. F. 1985. *Reflections on gender and science*. New Haven: Yale University Press.
– and H. Longino. 1996. *Feminism and science*. Oxford: Oxford University Press.
Kenway, J. and H. Modra. 1992. Feminist pedagogy and emancipatory possibilities. In *Feminisms and critical pedagogy*, eds. C. Luke and J. Gore. New York: Routledge.
Klein, N. 2000. *No logo*. Toronto: Vintage Canada.
Lather, P. 1991. *Getting smart*. New York: Routledge.
Lederman, M. and I. Bartsch. 2001. *The gender and science reader*. New York: Routledge.
Merchant, C. 1995. *Earthcare*. New York: Routledge.
Middleton, S. 1993. *Educating feminists: Life histories and pedagogy*. New York: Teachers College Press.
Milne, C. 1991. *Sacred places*. Saskatoon, SK: Western Producer Prairie Books.
Muzzin, L., P. Tripp, and E. Abergel. 2002. *Biology As If The World Mattered*, Presented at the 12th International Conference on Women in Engineering and Science, July, Ottawa, ON.
– P. Tripp, and P. Pennefather. 1997. Critiquing science. *In interdisciplinary perspectives: using qualitative methods to study social life*, ed. L. Muzzin, diskette, desktop publication, ISBN 0-9682062-0-4, Ottawa, Ontario: National Library.
Naess, A. 1989. *Ecology, community and lifestyle*. Cambridge: Cambridge University Press.
Orr, D. 1992. *Ecological literacy*. New York: State University of New York Press
– 1994. *Earth in mind*. Washington, DC: Island Press.
Plant, J., ed. 1989. *Healing the wounds: The promise of ecofeminism*. Toronto: Between the Lines.
Plumwood, V. 2002. *Environmental culture*. New York: Routledge.
Richer, S. and L. Weir. 1995. *Beyond political correctness: Toward the inclusive university*. Toronto: University of Toronto Press.
Roman, L. and L. Eyre. 1997. *Dangerous territories: Struggles for difference and equality in education*. New York: Routledge.
Schama, S. 1996. *Landscape and memory*. New York: Vintage Books.
Shiva, V. 1989. *Staying alive: Women, ecology and development*. London: Zed Books.
– 1993. *Monocultures of the mind*. London: Zed Books.
– 1997. *Biopiracy: The plunder of nature and knowledge*. Toronto: Between the Lines.
Shor, I. and P. Freire. 1987. *A pedagogy for liberation*. New York: Bergin and Garvey.
Simon, R. 1992. *Teaching against the grain*. New York: Bergin and Garvey.

Smith, D. 1987. *The everyday world as problematic: A feminist sociology.* Toronto: University of Toronto Press.
– 1990. *The conceptual practices of power. A feminist sociology of knowledge.* Toronto: University of Toronto Press.
– 1998. *Writing the social.* Toronto: The University of Toronto Press.
Smith, L. Tuhiwai 1999. *Decolonizing methodologies: Research and Indigenous peoples.* New York: Zed Books.
Teeple, G. 2000. *Globalization and the decline of social reform.* Aurora, Ontario: Garamond Press.
Wangari, E., B. Thomas-Slater and D. Rocheleau. 1996. Gendered visions for survival, In *Feminist political ecology: Global issues and local experience,* ed. D. Rocheleau, B. Thomas-Slater and E. Wangari, 127–154. New York: Routledge.
Wood, D. 2000. *Making Ontario.* Toronto: General Publishing Co.

SECTION ONE

Personal Experiences and Paradigm Shifts

1

The Seedlings Mattered

HEATHER MENZIES

First thing that spring I returned to the bit of savaged land we'd so virtuously reforested by hand, and saw nothing. None of the sprightly dark-green sprigs of spruce, or the stubby brush-fronds of pine that we'd planted. Nothing but last year's tall grass, thistles, and burdock weeds collapsed in a swoon from the drifting and soggy snows of a typical Ontario winter. I hunched down, my feet cold in my rubber boots. I took off my mitt, and lifted a limp strand of grass in my hand. Its colour had faded to a thin beige, with some liver spots of brown. Then I went digging, pulling the still-tough grasses apart where the thatch of weeds seemed to hummock a bit. And one by one I found them: the once lush and perky little seedlings my parents, brothers, sister, and I had dug and patted into the leached and stoney soil less than a year ago. Now they were bent, twisted, sometimes flattened completely, and all a sickly yellowing hue, or balded like a worn-down scrubbing brush. They looked about ready to die.

I looked around, then walked to a fencerow bordering this farm my parents had bought in a post-war bid for renewal and family traditions. The stones had been picked so hopefully out of these fields by whoever had cleared and tried to make a living off them. They were moss covered now and littered with branches of elms that had grown there and died. I collected a few grey branches and twigs, and headed back to the seedlings. I cleared the debris clinging to the single side branch of the one I'd first uncovered. I coaxed it into an upright position. Then, holding it up with one hand, I manoeuvred one of the dead-elm sticks into place to prop the poor little thing up.

It took a while to get the hang of it, digging the sticks into the ground, or wedging them into the tangle of weed and grass roots, then getting just the right amount of lean on the seedling so that it would both hold up the stick and be held up by it. I kept at it for hours, long past the sound of my mother yoo-hooing from the kitchen door and dishing out stew for dinner. I did the whole stretch before I stopped, each uncovering leading me to the next, each successful prop job readying me for another. Sometimes an elm stick was too rotted away and broke, so I had to go get another. Sometimes the seedlings slipped and flopped back prone against the ground, and I'd have to fine-tune the lean of the supporting sticks, the symbiotic crutch.

I was good and hungry when I headed home. Tired too, but happy as I looked back once more. And I can see them still, a rag-tag array of pewtered, crooked dead-elm sticks, with patchy, pale-needled seedlings draped awkwardly against them.

Looking back now, they mark my entry into another way of understanding, of knowing the world – as if it mattered. I began to know and understand the world as it is, on its own particular terms, as these caught my attention that April day. It was a threshold to a journey, as surprising and unexpected as the wardrobe conduit into Narnia, in a favourite book my teacher was reading out loud at the time. It was just as hidden too, coming at least a year after I would have expected my journey to begin.

The previous spring, my ten-year-old self had taken up the shovel like a torch, onward to the uplands of this over-worked, then abandoned, little farm, to plant 10,000 government-issue seedlings in a reforestation project my father had signed on to after soil samples showed the ground too leached and eroded from exposure for crops. I'd understood about the resources of nature, and how they can be depleted, soil being one of them. To reverse the process, you put things back the way they were. In this case, by replanting the trees that should never have been cut down in this part of Ontario. The pre-Cambrian shield runs like a line of old dinosaur bones beneath the land here. What passes for soil is a thin skin of half-digested glacial morraine, making it a poor prospect for growing much of anything, even trees. The shock of returning to the site of all our confident replanting to find imminent disaster taught me that it isn't quite as simple as that. Life is not a laboratory experiment that can be repeated at will. It's last year's tall grasses and dead thistles proving as lethal as clear-cutting corporate interests. It's one damn thing after another. On that raw Spring day when I was 11, I learned that life exists in context: a living, interwoven matrix of space and time. Things can't be plucked out of one context and stuck into a new one just like that. It takes time to nurture something living into place, within the particulars of its context, even a mundane one made up of overgrown weeds.

Time dwells in the border regions between things. Time works with their particular affinities, inclinations, and intentions, conjuring relationships and context, the radical interdependency that is life. The truth of this can be found in science, as well as culture. I think of a botanist colleague at Carleton University, Dr Margaret McCully, who studied soil sheaths that grow around the roots of field corn. She learned that they are a mixture of root cells, soil bacteria, and soil particles held together by a kind of mucus secreted by the root (Menzies 1991, 32). In making this discovery, Dr McCully offered an illustration of what seems to be emerging as a crucial modification of Darwin's theories of evolution. Organisms don't blindly adapt to their environments or die. As Richard Lewontin put it in the 1990 CBC Massey Lectures, *Biology as Ideology*, organisms interact with the world around them to create the environment they need. It's worth quoting Lewontin at some length on this point, to understand its implications. "When we free ourselves of the ideological bias of atomism and reductionism and look squarely at the actual relations between organisms and the world around them, we find a much richer set of relations, relations that have very different consequences for social and political action. First, there is no 'environment' in some independent and abstract sense. Just as there is no organism without an environment, there is no environment without an organism. *Organisms do not experience environments. They create them. They construct their own environments out of the bits and pieces of the physical and biological world and they do so by their own activities*" (Emphasis added. Lewontin 1991, 83).

We gather wood and build heated houses to survive Canadian winters. I gathered bits of dead elm trees to help those seedlings survive. Corn plants construct soil sheaths. To hear Dr McCully describe it, it's almost a pact between the root hairs "working hard" as she puts it, to collect the nutrients they need from the soil, and enlisting help from the "friends" they have in the soil microbes, who unlock the nutrients in the soil so the root hairs can absorb them. It's a complex and not at all haphazard symbiosis.

Lynn Margulis built on her pioneering work studying genetic material outside the chromosome to articulate an entirely new paradigm of evolution based on this principle. Specifically, she claimed that evolution hinges on a set of symbiotic relations between microbes that created an environment in which more complex forms of life could emerge, through symbiogenesis (Margulis 1998). Life is not just "live and let live" among organisms that border each other, she argues. It's a mutual enabling of life – "letting live," not passively but actively, through mutual engagement, even dialogue.

What emerged through a modern scientist's microscope – a world view of mutual implication and symbiosis – resonates sweetly with some very old knowledge traditions: those associated with Aboriginal cultures. In Canada, I interviewed an Algonquin Elder, who called herself "Kokom" or grandmother, for a video I was making about Native approaches to science and technology. Kokum talked about the complex interdependencies that constitute our lives. "In depending upon each other, we want to wish for one another to co-exist," she told me, through her grandson Jacob Wawatie, acting as interpreter. "For example, if you don't wish for the blueberries, you don't pick them in the summertime. Then the spirit of the blueberry will not come back. So you must keep on using the land to co-exist. And you do it for all the medicines to keep on existing. Because if we stop using it, what are the reasons why they want to keep on existing, existing themselves, the spirits? So the spirits leave. The medicine in the medicine plants is not strong anymore, because we don't believe in it" (Menzies 1997, 49).

There is a similar interconnection between knowledge and engaged social practice in the knowledge systems of the Yolgnu people of Australia — at least from the glimpses I've gained through the work of Helen Verran. Dr Verran, a philosopher of science at the University of Melbourne, Australia, was invited into this tribe so that she could interpret its traditions to the Western world. The Yolgnu's knowledge tradition is called "gurrutu" (Wertheim 1995, 39). The model for depicting it, she discovered, is the ten fingers of the hands. The fingers serve as a metaphoric framework from which the whole system, the "gurrutu mesh," of understanding and knowledge about the social and natural environment hangs, and from which each person and family's responsibility flows. That's because the finger-counting network is first a network of family relationships, enumerated through three generations. The connection to the land and related knowledge flows from there. Each of the kinship relationships in turn serves as a point of connection to a network of relationships to the land, and to knowledge of how to co-exist with the land – with much of the latter embedded in "songlines," part creation myths in which Australians' totemic ancestors created a trail that, when followed and sung, helps keep the land alive (Chatwin 1987, 12 and 52). So the fingers depict almost the opposite of what the digits Western knowledge has derived from them depict – single units in isolation. For the Yolngu, they are points of reference in a network of relationships, and simultaneously a way of knowing the world through kinship relationships that actively places each individual inside the picture, inside that network of social relations and knowledge. According to Verran, the gurrutu mesh "is like a very complex locating system" in which "the world comes alive already mapped."

I think back to my 11-year-old self trying to take some after-the-fact responsibility for land that had been stripped of its Aboriginal matrix of songs and myths coding knowledge for custodial renewal by its first peoples. The land was stripped of all that when it was appropriated into European maps, where it became real estate that changed hands, first from the British crown into a clergy reserve, then to the MacMillans and the Viaus and then, after the war, to us, the Menzies family. I think of my naïveté, my fumbling attempts to improvise and fill the sterile gap – the starved lack of knowledge epitomized by the leached and sterile soil. Still, as my eye returns to that weed-pocked old field studded with bent grey sticks and crippled spruce and pine seedlings, I smile. I learned something. I did something, and it mattered. Then I realize something else. What I learned and came to understand through my efforts is not just about relationships and their importance. It's *in* relationships. I learned what I did in those relationships, and through the medium of those relationships. It's epistemology as engaged relationships: I learned what I learned through being involved and engaged, with knowledge and insight emerging precisely because of my becoming involved. What I was able to see and sense and know might easily have escaped the notice of the detached, outside observer.

When did I get involved? Nominally, the previous spring, in what I thought was a one-shot deal. It was a chore really, a grudging going along with a project set in motion by my father, getting us all to work on things together. I'm sure I whimpered enough. I recall even breaking down and crying more than once, in frustration and exhaustion from hours and hours of trying to shove my shovel into that stoney, inhospitable ground. Then I'd root around in the still ice-crystalled earth with my bare hands, pulling out stones that would block the roots. Dirt rammed under my fingernails as I dug with cold-reddened fingers, scrabbling together enough gravel to cover the hungry roots. My fingers ached with the cold, and were swollen fat as sausages. On good days, at least I dressed up the work in heroics – saving the God-forsaken land and all that. Yet somewhere, on the other side of the tears, over the long haul of those weekends planting trees, something more subtle and personal must have taken hold. Something drew me back the following Spring. Some filaments of connection, some affinity and fellow feeling had crept through the pores of my water-soaked, cold-stiffened hands. Hardly a full-blown gurrutu matrix of connection and responsibility. Still, it was enough that it hooked me in that following spring, and kept me there past dinner time, past even hunger and fatigue, learning, improvising, slowly getting it right. I learned what I did in those relations, in the context of that involvement, and because of it. Somehow I felt implicated; the fate of these trees mattered.

So for me, it's not incidental to Marg McCully's discoveries about soil sheaths that she was raised on a farm full of corn fields, and as a scientist she wanted to take her Harvard PhD in cell biology back home in some way, "and apply what I knew to the real thing, corn in the field." Ironically, soil sheaths do not show up on laboratory corn, or they are washed off in the protocols for preparing the samples. This might explain how their initial "discovery" (in 1882) came to disappear, why this bit of knowledge became lost to twentieth-century science. Dr McCully discovered soil sheaths afresh because she chose to study corn in the naturalists' tradition, through observation in the field, studying them on their own terms, not on her own or on laboratory-prescribed ones. She came to know what she did because of that field experience, because that's where her work mattered: in real corn in real fields tilled by real farmers.

Fieldwork relations played a similar role in Dr Barbara McClintock's breakthrough theorizing about genetic transposition in the 1920s. From her field observations, she concluded that genetic mutations in corn were anything but blind and random, as genetic orthodoxy then maintained. Rather, they seemed to emerge as part of an adaptive regulatory system in which the corn plant responds to even small environmental stresses through minute transpositions of genetic material, which in turn help keep the plant in a state of "dynamic equilibrium" (Fox Keller, 1983, 193). Dr McClintock spent six years struggling to articulate this theory, sustained, as she said, by the joy of her relationship with the plants she worked with – by her "thorough absorption in, even identification with, her material," as her biographer, Evelyn Fox Keller, put it. She waited another 30 years before the molecular microscope helped to confirm the validity of what she had observed, by keeping her eyes and mind open to how corn plants seemed to interact with their environment.

When people asked her how she had the courage, the imagination, the inspiration to conceive of genetics as a kind of dialogue with the environment, her answer always touched on her relationship to what she studied and sought to know. Over and over again, she tells us one must have the time to look, the patience to "hear what the material has to say to you," the openness to "let it come to you." Above all one must have "a feeling for the organism" – an attuned sensibility that McClintock found helped to "extend her vision" (Fox Keller 1983, 198). Fox Keller wrote the biography of Dr McClintock with the conviction that her subject's medium – her attuned dialogue-like relationship with the corn plants – was an inalienable part of her message, and as important to the world as the knowledge that garnered her a Nobel prize in the end.

I'm back with that rag-tag collection of half-rotted sticks and winter-weakened seedlings. Our relationship was less a dialogue, I think, than a dance, a dance of survival. I remember my young hands reaching out, reaching for each one of them in turn, lifting them up, providing a supporting arm – through the stick. They were a tottering lot, good for a slow shuffle at best, that first year. Correction: *we* were a tottering lot, then and the following spring and longer. Yet now I can dance among them — and love to do so when a fresh spring wind comes whiffling through their slow-swaying thick-needled branches. Forty years later, they are a full-fledged stand of mature, and largely straight, trees, mulching the now replenished ground with needles and a few dead branches of their own.

Walking among them, brushing my fingers through their glistening fronds, sniffing the pine gum and the fecund richness of moist earth, I pull yet another lesson from my experience. It wasn't just how much I learned through my relationship with those trees. It's how I expressed it that mattered. In a way, I guess it was a dialogue, beginning with my being called back, and then compelled somehow to respond to the seedlings' winter-flattened plight. It was my groping bare hands that made a difference. My hands improvising habitat, learning from the seedlings themselves (what manner of propping they needed, what angle of incline of stick to young shoot.) I helped them create the necessary environment in which to recover from their unnatural exposure to that first winter in an open field, and to survive the next one. I helped to articulate, to name, the story of their eventual adaptation to that environment, not through words but through action. It's a bit like the story-myth the Algonquin elder Kokum told me about the spirit of the blueberry needing some creature to reach out hungry for the plants to keep producing juicy fat blueberries every year. That same reciprocity and mutual affirmation is what matters, in the context of those relationships. Then I remember that the ongoing dialogue between an apprentice-student and a story-telling Elder, and within that between the student and the story itself, is the pillar around which much First-Nations teaching is organized. I begin to suspect that this might be a key to teaching as if the world matters.

The Russian philosopher Mikhail Bakhtin spent a lifetime theorizing about language, literature, and relativity, and came to the provocative conclusion that human existence is a participatory event articulated not just in dialogue but *as* dialogue. "The word is born in a dialogue as a living rejoinder within it ... Forming itself in an atmosphere of the already spoken, the word is at the same time determined by that which has not yet been said but which is needed and in fact anticipated by the answering word" (Bakhtin 1981, 279–80).

Bakhtin's is a rich and elaborate body of thinking, and I can only draw selectively from his key concepts here to help bring home how his point applies to teaching – as relationships, as dialogue. One concept is that we are all embedded in life, first and foremost in our bodies. Another is that life addresses us, and calls us to respond, expressing the particulars of our being, however and wherever we are situated in life's vast interconnected matrix of time and space. Far from being alone, he argues that our existence is always one of "co-being." When we speak and otherwise utter (or "outer") and manifest our existence, he argues, we can never do this as some disembodied, detached observing "I." We are always and inescapably participants, and we always articulate that in dialogue. The dialogue is first an internal one, with all that remains unspoken within us as inarticulate dreams and memories. It's only secondly an external dialogue, anticipating another person in the choice of words, intonation, and cadence that informs our speaking, toward another.

In some excellent work interpreting Bakhtin, Michael Holquist emphasizes the influence of Einstein's theory of relativity (the observer ineluctibly implicated as participant), and concludes that Bakhtin essentially interpreted relativity as epistemology – this epistemology being dialogism (Holquist 1990, 20). In other words, the best and truest way of knowing the world is in the relativity of communicative relationships, as dialogue. While others used relativity to rethink science as a way of knowing the world, Bakhtin applied the concept to literature and culture. He was particularly fascinated by the novel, which to him articulated this embodied, embedded way of expressing the truth of the world.

Word by word, molecule by molecule, moment by moment, a story is articulated. It's a life line, the story of life as how we work out the stuff that matters. That's the key. It doesn't lie in an isolated abstraction called the WORD. It's in the spoken word and the listening ear, in relationships of dialogue. It's the voice of a living being attuned to life as it calls to us, in the particulars of space and time. It's our responding, saying YES.

Teaching as activism is all rather simple really. We help create a world where all life and all people matter – even, as Ursula Franklin once put it, where "all people matter equally" (Franklin 1984, 11) by acting on that belief, that conviction. In other words, by modelling such a world in the classroom, through our own relationships with students and through the relationships of learning and self-expression, we encourage as we entice our students to be implicated participants themselves. All the best teachers I've ever had did this – starting with the one in grade six who paid attention to the pale, scrawny seedling of a kid hunched over somewhere near the back of the classroom and, by taking the time to draw me into class activities, made me feel that I mattered. That's the year I finally learned to read.

REFERENCES

Bakhtin, M.M. 1981. *The dialogic imagination*. Austin: The University of Texas Press.

Chatwin, B. 1987. *Songlines*. New York: Penguin Books.

Fox Keller, E. 1983. *A Feeling for the organism: The life and work of Barbara McClintock*. New York: W.H. Freeman and Co.

Franklin, U.M. 1984. *New approaches to understanding technology*. Technology, Innovation and Social Change: Conference Proceedings. Ottawa: Carleton University (School of Canadian Studies).

Holquist, M. 1990. *Dialogism: Bakhtin and his world*. London: Routledge.

Lewontin, R. 1991. *Biology as ideology: The doctrine of DNA*. Toronto: Anansi Press.

Margulis, L. 1998. *Symbiotic planet: A new view of evolution*. New York: Basic Books.

Menzies, H. 1991. Science through her looking glass. *Women's education des femmes*. Vol.9, #2 (Fall).

Menzies, H. 1997. *Canada in the global village*. Montreal: McGill-Queens University Press.

Wertheim, M. 1995. The way of logic. *New Scientist*, 2 December.

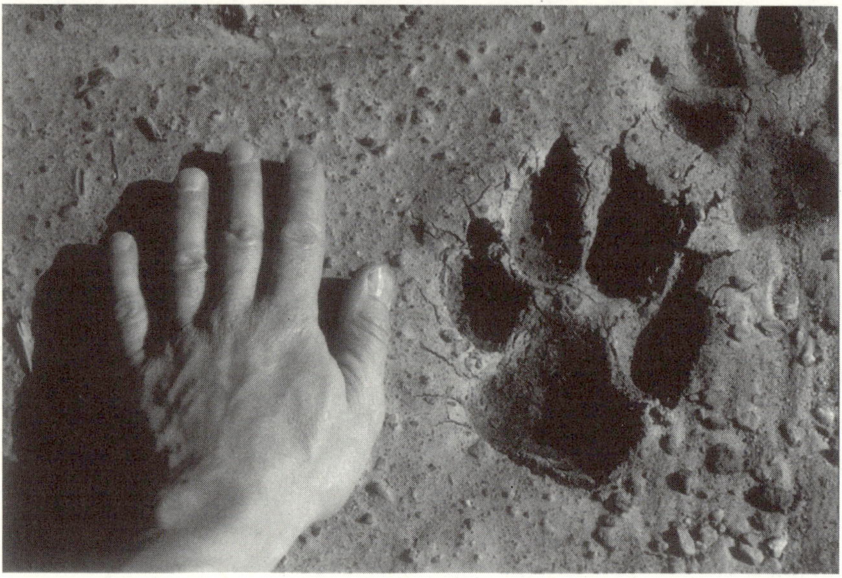

2

"The Wolf Must Not Be Made a Fool Of": Reflections on Education, Ethics, and Epistemology

BOB JICKLING

A WOLF STORY

Last summer a group of canoeists pulled into an eddy, then stepped out of their boats onto the rocky edge of a dry alluvial fan. Aqua-tinted water of Yukon's Wind River drifted by. Inquiries about stopping – again, so soon – shifted to the white figures on the opposite bank. With attention, these figures gradually took the shapes of Dall sheep at a mineral lick.

Seeing sheep at a mineral lick isn't rare – not an everyday experience, but common enough if you know where to look. Journeying by canoe allows travellers to slip, for a time, inside a corner of their ecosystem. But this day there was a black figure too. Head down, it popped out of foliage at the river's edge, ambling, searching, and occasionally breaking into a trot. Responding to some stimulus, it stopped. Alert, its ears grew, strained, and focused; trotting turned to cantering. Then it was back to trotting, nose down and weaving through the willows. As it stopped again, a little closer, clues of gesture, carriage, and shape whispered "wolf."

The black wolf's nose dropped toward the ground then popped up again. Up and down several times, it seemed to be sizing up the visitors.

"Do you suppose she thinks we're prey?" Interesting question.

Does this attention make the travellers, or the next sheep, objects for the wolf – potential nourishment for the pack? Does this make them voyeurs in a bifurcated world of eater and eaten, predator and prey, human and non-human?

No. This time the visitors were subjects of curiosity, and purveyors of curiosity, watching yet not staring, as they shared a few moments with the residents.

ANOTHER WOLF STORY

Systematic culling of Yukon wolves is not new. Rather, killing of wolves to benefit humans is part of our contemporary Canadian history. The first biologist hired for the Yukon in the 1950s was assigned the task of overseeing a widespread wolf-poisoning program (Gilbert, 1994). At the time, public norms had relegated wolves to the status of vermin. By contemporary standards, the scale and methods of this program are unthinkable. Nevertheless, modern wildlife management has rested on assumptions that are fundamentally the same as those of the Yukon's pioneer biologists.

"Wolf management" is controversial in the Yukon. In 1997, the government completed a five-year control program that entailed the killing of 80 percent of the wolves in a 20,000-square-kilometre region of southwest Yukon. The killing was accomplished by shooting wolves from helicopters or, in later years, strangling them in snares. The justification for culling was concern for declining numbers of caribou in particularly vulnerable herds. In this case, the objective was to prevent the Aishihik caribou herd from declining below a level where a natural recovery would not occur quickly enough. Because caribou are important for subsistence hunting, the government of the day judged that a long-term natural recovery, estimated to be between twenty and thirty years, would not be acceptable to local people (Renewable Resources, 1995).

Not everyone supports wolf kills and many spoke out in opposition. They did so for a variety of reasons; environmental issues are rarely simple. Whereas some Yukon voices said that it was important to rebuild the Aishihik caribou herd, others were concerned that too little was known about why this herd was declining. Others claimed the wolf was being made the scapegoat for past excesses and that over-hunting was responsible for the problem. For many, the critical question was whether the wolf kill was a reaction to a biological problem or a treatment for the symptoms of a much deeper human problem.

In the later stages of the Aishihik wolf-culling program, biologists began experimenting with wolf sterilization techniques as an alternative to shooting. The method involved a) tranquillizing wolves from a helicopter; b) moving the wolves to a nearby community and placing them in a holding crate; c) on arrival of a veterinarian, anesthetizing the wolves then sterilizing them with either a tubal ligation or a vasectomy; d) placing wolves in release boxes for the night; and e) returning them to the capture site the next day. In an effective public relations initiative, local managers invited a newspaper reporter to observe their fieldwork. In a three-page spread, the journalist (Tobin 1997) reported that "wildlife managers" proposed sterilization of the dominant males and females in wolf packs. The theory is that non-breeding pairs will defend their territory but in the absence of a growing pack will consume less moose and caribou, particularly caribou calves.

The journalist reported that Yukon's wolf sterilization program represented a "world-wide lead into a new area of wildlife management," and that these techniques will "replace the more 'intrusive' aerial hunting" and respond to "the public's growing disdain" for this technique. Finally, the article reported a biologist's prediction that "sterilization will reduce the frequency in which aerial hunting must be carried out. That will soften public resistance, she feels, and be less of a drain on financial resources" (Tobin 1997, 23–25).

A fundamental problem reflected in the above comments is that many biologists do not recognize that all knowledge has an ethical component. Immersed in "wildlife management," the biologists have declared their bias. Creating an illusion of thoughtfulness, choice, and objectivity, their prejudices nevertheless are revealed. Presenting the public with a choice of more or less intrusive techniques does nothing to challenge, or even problematize, industrial societies' drive for the ever-more-effective domination of nature and use of its "resources." Many opponents of wolf sterilization do not want less intrusive techniques – they want new ethics. An agency organized for maximum sustainable harvest is very different from an agency organized for conservation.

In a recent development, the Yukon government has approved the Aishihik Integrated Wildlife Management Plan (AIWMP 2000) that calls for ongoing wolf suppression through use of sterilization techniques. Implicit in this plan is the prospect of long-term wolf control.

SCIENCE ISN'T OBJECTIVE

The hard reality revealed in the Yukon example is that science is not objective. To be fair, it is not completely subjective either. We sometimes create untenable discussions when we allow distinctions to become dichotomies. Perhaps a useful way of seeing this distinction is as representing binary opposites on a continuum from absolute subjectivity to absolute objectivity. All science would lie somewhere in the middle. Principles of fairness, falsification, and blind peer-review do help science in providing snapshots of reality. However, the main point here is that it only provides small snapshots; much remains unknown and uncertain. Often our ethical choices are framed and driven by what we know. In the Yukon example, attempts are made to bind public discussion to culling practices that are "more" or "less" intrusive. Even traditional animal welfare arguments are based on knowledge claims, often about sentience. In essence, we most frequently operate within frameworks of epistemologically based ethics (Cheney and Weston 1999). If science is not ultimately objective, then it is important to understand how, intentionally or implicitly, science can limit available ethical choices. Consider again the Yukon wolf kill program.

Biologists responsible for "wildlife management" have been central to the public discourse on the killing of wolves. While their task is to enact public policy, they too hold assumptions that shape the way policy decisions are presented to the public and subsequently implemented. They also shape their image through strategic public relations campaigns.

With an insightful comment made during the planning for the Aishihik wolf kill, one biologist said, "There are two options: intensively manage or let it follow a natural decline. Both are defendable" (Farnell 1992, 8). From a scientific perspective, two defensible alternatives were seen as possible: intervene in the ecosystem by removing wolves and monitor the effects, or allow the system to decline and/or recover without human intervention and again monitor the effects. The same biologist went on to say that the "[p]ublic in [the] Yukon prefers intensive management on this herd" (1992, 8). A colleague added, "In Aishihik, we need to address public policy ... and design something scientific, with a set of hypotheses and alternate hypotheses" (Hayes 1992, 8).

Although the extent to which people in the Yukon prefer intensive management is debatable, a level of public support for a wolf kill existed. The government of the day and some biologists were responding to a vocal segment of the population. However, to say those managers simply responded to public will is to underestimate the complexity of the relationship. Trained as wildlife biologists, their enthusiasm for science is easy to see. Steeped in traditions rooted in Bacon's seventeenth-century creed – "the secrets of nature reveal themselves more readily under the vexations of art than when they go their own way" (1960, 95) – there was ready approval and acceptance of experimental intervention among managers who often feel the need to manage.

Nevertheless, as noted above, it is far from clear that the new science of wildlife management rests on assumptions that are fundamentally different from those of the Yukon's pioneer biologists. The killing of wolves is now dressed with the "respectability" of science and the "objectivity" of experimental design. Unfortunately, this presumed respectability masks questionable philosophical underpinnings. It is notable, for example, that there was no further consideration of the "hands off" option of monitoring the natural decline of Aishihik caribou and wolf populations. Somehow, we seldom conduct this experiment.[1] We might also ask what "ethical" choices are available for the public and its decision makers when our ecosystem snapshot is based on observing a "vexed" nature, a nature subjected to intrusive "treatment" conditions. Alternatively, what choices would we have if our inquiries had begun in the manner of those wolf observers described at the beginning of this discussion? These observers watched unobtrusively. The wolves were subjects of their respect, interest, and curiosity, not objects of science, management control, or human domination.

Of course casually watching wolves, or any other species, from a riverbank is not enough to make choices or to resolve difficult disputes, but it does illustrate a basis for a different philosophical framework – and this is the key point. Choosing to observe animals unobtrusively, or to monitor population cycles without human intervention rests on particular assumptions and values. And these values are different from assumptions that presume moral and political authority to manage and manipulate wildlife populations and impose vexatious treatment conditions. Clearly, the knowledge we have to work with depends on the ways we approach the world and the kinds of questions we ask. As Cheney and Weston (1999) point out, the world is not completely knowable; we can only know portions based on the kinds of inquiries we pursue. Ultimately, these questions are rooted in our values – our ethics – making all knowledge value-loaded and ethics-based.

TOWARDS AN ETHICS-BASED EPISTEMOLOGY

Our epistemologies, our systems of knowledge, rest on ethical choices whether these are made consciously or not. Epistemology is value-driven. This being the case, then an *a priori* concern for all scientists is to consider how they ought to approach the world in their pursuit of knowledge; what is an ethical approach to science? This is a radical shift for ethics too. Instead of looking for knowledge claims to frame ethical discourse, I argue that ethics are primary; they open the way to knowledge.

In posing concern for ethics, we are trying to discover what things in the world demand practical respect. Building on the work of Tom Birch (1993) and Cheney and Weston (1999), I suggest that this is an open-ended question that demands open-ended, nonexclusive consideration of everything, insofar as this can be done. Birch calls this "universal consideration," whereby "[o]thers are now taken as valuable, even though we may not yet know how or why, until they are proved otherwise" (1993, 328). Such a view carries obligations. As Cheney and Weston (1999) argue, "universal consideration requires us not merely to extend this kind of benefit of the doubt but actively to take up the case, so to speak, for beings so far excluded or devalued" (1999, 120). Fundamental to this requirement will be seeking to understand how we ought to approach those things that we respect.

That we should approach other entities with respect and that knowledge is ethics-based are not new ideas. Many traditional ways of knowing have stood in contrast to the epistemological dominance of Western science, including those of Yukon First Nations.[2] I cannot interpret First Nations cultural traditions; they will have to do that for themselves. I can only comment on my own experiences while listening to, and working with, Aboriginal colleagues. The legends, stories, and reflections of Yukon First Nations people have placed before us a mirror which offers an opportunity to reflect upon our own cultural traditions. Of particular interest is the challenge to our culture's framework for organizing knowledge – our tendency to separate ethical, emotional, and spiritual knowledge from "hard" science.

The Western approach to knowledge – separated into component parts and assigned to different disciplines – can be contrasted with traditional modes of knowing in which the ethical dimension is given its due emphasis. This contrast is revealed in Native Heritage Advisor Louise Profeit-LeBlanc's response to a question about "truth" in her peoples' stories (1996). Here she used the Northern Tutchone term *tle an oh* (klee-ah-no). A difficult term to translate, it is said to mean something like "correctly true," "responsibly true," "true to what you believe in," "what is good for you and the community," and what "rings true for everybody's well being" (in Cheney 1999, 151). Here we have evidence of an epistemological framework that has an inherently ethical dimension. Put another way by Carol Geddes, "We would never have a subject called environmental ethics; it is simply part of the story" (1996, 32). The argument for an ethics-based epistemology seems to be an attempt to recover an understanding that has never been lost in many traditional cultures.

If all knowledge is ethics-based, and if, in the absence of evidence to the contrary, all entities deserve consideration, then how ought we approach inquiries about these entities that we respect? First, the idea of universal ethical consideration begins to create new ways of seeing the world. How we create or recreate the world counts. We can no longer be aloof or disinterested observers. Ethics-based epistemologies are concerned with right relationships. And right relationships are grounded in mindfulness. When we are mindful and respectful then we act with courtesy and etiquette, including trans-human etiquette. Louise Profeit-Leblanc underscores this point when asked to comment on ethics. For her, ethics are "that which we do to ennoble us" (1996, 14). Ethics then are more than collections of ideas; they are also performative.

What we do, how we act, and the research procedures we choose all count. Aldo Leopold, often thought of as the father of wildlife management, knew this too. In his famous essay, "Thinking Like a Mountain," Leopold recounts a life-changing experience that occurred on the day he saw a wolf die: "We reached the old wolf in time to watch a fierce green fire dying in her eyes. I realized then, and have known ever since, that there was something new to me in those eyes – something known only to her and to the mountain. I was young then, and full of trigger-itch; I thought that because fewer wolves meant more deer, that no wolves would mean hunters' paradise. But after seeing the green fire die, I sensed that neither the wolf nor the mountain agreed with such a view" (1970, 138–139). For Leopold, shooting that wolf was not ennobling and he felt this deeply.

Returning to the Yukon for a moment, I find it difficult to imagine how a helicopter-assisted killing of 80 percent of the wolves in a 20,000-square-kilometre region can be considered respectful or sterilization of dominant pairs in a wolf pack considerate. People, particularly scientists, are not eager to talk about this. However, Tlingit Elder Harry Morris, commenting at a meeting before the Aishihik wolf kill, provided those willing to listen with much to think about. He said that his people kill wolves, they don't mind killing wolves, but "the wolf must not be made a fool of" (1992). For Mr Morris, right relationships and right conduct matter.

IMPLICATIONS FOR EDUCATION

Science does not exist apart from other ways of understanding the world. It is not objectively aloof and disinterested in the world. And science is not benign; we all live with the consequences of scientific activity. Development of the atomic bomb had ethical dimensions aside from the "good science" employed to accomplish the task. Unresolved moral questions continue to frame the debate over use of the bomb on Hiroshima and Nagasaki. And these points have implications for how we conceive of science in education.

Environmental philosopher Arne Naess describes how children learn that scientific knowledge is "something opposed to myths, and the undue influence of feelings, and values." He goes on to suggest that "you easily get to overestimate the importance of scientific knowledge in a vital question, which is always also a value question" (Naess and Jickling 2000, 55). As presented, science often denies its ethical basis and its moral responsibility. For Naess, we need to undermine not science, but the prestige of science, in favour of better understanding value priorities. Another way of framing this observation would be to find ways to build up an understanding of ethics – to remove barriers to its practice and to make ethical dimensions of science more explicit.

A starting point for removing barriers to understanding the ethics that underlie science would be to be more up front about the limits of science. Science is not, after all, about truth. At best it is about probabilities. And these probabilities are often validated through processes such as blind reviews in which the outcomes can be affected by how these reviews are conducted and who is chosen to provide them. We should teach students about the probabilistic nature of science, about the review processes, and about the need to be vigilant about how these processes are conducted.

Next, science rests on assumptions; Thomas Kuhn (1970) made this much clear in his classic book *The Structure of Scientific Revolutions*; others have followed in his wake. Education can enable students to explore those assumptions by asking "What kinds of questions are being asked?" "Why?" and "How do the scientists seem to be approaching the world – as an object of vexation, or as a subject of respect?" Answering these kinds of questions can enable us to be more conscious of the science that we need – and that we believe is more ethically justified.

Both science and education have performative dimensions – what you do counts. It will be difficult to convince the public that your science rests on an ethic of respect or etiquette if it follows the Baconian ideal of vexing nature – if it entails, for example, killing large numbers of wolves. Similarly, it is difficult to imagine a respectful classroom, grounded in universal consideration or trans-human etiquette, where students never see, touch, smell, or listen to other living beings – or worse, classrooms where dissections are commonplace. Educators, too, can do that which ennobles their instruction.

Ethical approaches are not simply objective; there is always a more subjective and emotional component. We need to pay attention to this. As Arne Naess (2000) says, "There is an underestimation of the cognitive values of feelings" (Naess and Jickling 2000, 53). Following Naess, we should allow space in our instructional programs for his sequences of questions: "How do you feel?" "What do you feel?" Then, "What should you feel?" "What do you think you are right to feel?" and "What do you want yourself to feel?" (Naess and Jickling 2000, 56). Naess's views are nicely complemented by Val Plumwood's (1999, 75) observation that "[m]oral reasoning requires some version of empathy, putting ourselves in the other's place, seeing the world to some degree from the perspective of the other with needs and experiences both similar to and different from our own." Building our ethical understanding will, therefore, be linked to our ability to develop emotional understanding. It is here that etiquette and respect are grounded.

Ethics, unlike science, is not a problem to be solved. To many, it is characterized as dynamic, evolving, and pluralistic. We are thus not seeking right answers, but engaging our students in a process that can help them to make better decisions. But, it is a process that requires ongoing reflection and revision. There is much that individual scientists and educators can do, but environmental ethics is a broad new field of inquiry and scientists should also be encouraged to build alliances with philosophers and ethicists. Such collaboration recognizes the interdisciplinary nature of environmental issues, and all these fields will be enriched by this work.

Finally, ethics should be an everyday activity – a normal activity (see Saul 2001). If we do not place science in its ethical context, then we run the risk of saying that ethics is not important or that it is reserved for experts or heroes who reside elsewhere. Given the enormous prestige accorded science in our society, and the responsibility that scientists and beneficiaries of science hold, we must reunite science with its ethical underpinnings to make our stories whole again.

NOTES

1 Currently there is an example of such an alternative "experiment" in the Yukon. Grounded in a local initiative, members of the Carcross and Tagish communities have worked with Rick Farnell and other biologists to implement the Southern Lakes Caribou Recovery Project (O'Donoghue 1996; Farnell et al.1998). This program assumes that humans, not wild carnivores, must be managed. Indications suggest significant increases in caribou populations without predator controls.

2 It is easy for cynics to point to questionable environmental practices – perhaps indiscretions – of First Nation individuals and communities to undermine their credibility. However, we are in danger of confusing moral inconsistencies with the moral possibilities contained in rich cultural traditions. We are all inconsistent. As Arne Naess recently said (Naess and Jickling 2000, 58), "It's a high ideal to be consistent. And, you will achieve it when you die – not before." We might look past these perceived inconsistencies and reach across this cultural divide; and, with our Native American colleagues, create new possibilities for environmental thinking and etiquette. Perhaps a little generosity of the heart and the head is needed.

REFERENCES

Aishihik integrated wildlife management plan (AIWMP). 2000. Whitehorse: Yukon Renewable Resources, Champagne and Aishihik First Nations, Alsek Renewable Resource Council (see also http://www.yfwmb.yk.ca/comanagement [accessed December 4, 2003]).

Bacon. F. 1960. *The new organon*, Book 1, Aphorism XCVIII, In *The new organon and related writings,* ed. F. H. Anderson, 94–95. Indianapolis: The Bobbs Merril Company. (First published in 1620).

Birch, T.H. 1993. Moral considerability and universal consideration. *Environmental Ethics, 15*(4), 313–332.

Cheney, J. 1999. The journey home. In *An invitation to environmental philosophy,* ed. A. Weston, 141–167. New York: Oxford University Press.

Cheney, J. and Weston A. 1999. Environmental ethics as environmental etiquette: Toward an ethics-based epistemology. *Environmental Ethics, 21*(2): 115–134.

Farnell, R., R. Florkiewicz, G. Kuzyk, and K. Egli. 1998. The status of *Rangifer tarandus caribou* in Yukon, Canada. *Rangifer* (10): 131–137.

Farnell, R. 1992. In *Designing an experiment for large mammal recovery in the Aishihik area, Yukon Territory*. Minutes of technical meeting, October 4, 1992, 8. Whitehorse: Government of the Yukon.

Geddes, C. 1996. What is a good way to teach children and young adults to respect the land? A panel discussion. In *A colloquium on environment, ethics, and education*, ed. B. Jickling, 32–48. Whitehorse: Yukon College.

Gilbert, S. 1994. Science, ethics and ecosystems. In *Northern protected areas and wilderness*, eds. J. Peepre and B. Jickling, 195–201. Whitehorse: Canadian Parks and Wilderness Society and Yukon College.

Hayes, B. 1992. In *Designing an experiment for large mammal recovery in the Aishihik area, Yukon Territory*. Minutes of technical meeting, October 4, 1992, 8. Whitehorse: Government of the Yukon.

Kuhn, T. 1970. *The structure of scientific revolutions, 2nd Ed*. Chicago: University of Chicago Press.

Leopold, A. 1970. *A sand county almanac*. New York: Ballantine Books. (First published in 1949).

Morris, H. 1992. Oral comments presented to the Yukon Wolf Management Planning Team, April 11. Whitehorse, Yukon.

Naess, A. and Jickling, B. 2000. Deep ecology and education: A conversation with Arne Naess. *Canadian Journal of Environmental Education* (5): 48–62.

O'Donoghue, M. 1996. *Southern Lakes caribou recovery program*. Progress Report 1992–96. Whitehorse: Council of Yukon First Nations.

Plumwood, V. 1999. Paths beyond human-centeredness. In *An invitation to environmental philosophy*, ed. A. Weston, 69–105. New York: Oxford University Press.

Profeit-LeBlanc, L. 1996. Transferring wisdom through storytelling. In *A colloquium on environment, ethics, and education*, ed. B. Jickling, 14–19. Whitehorse: Yukon College.

Renewable Resources, Government of the Yukon. 1995. The Aishihik program. In *Yukon wolves: Ecology and management issues*, ed. C. Olsen, 13–14. Whitehorse: Yukon Conservation Society.

Saul, J.R. 2001. *On equilibrium*. Toronto: Penguin.

Tobin, C. 1997, February 14. Lessening the bite ... into nature's meat supply. *The Whitehorse Star*, 23–5.

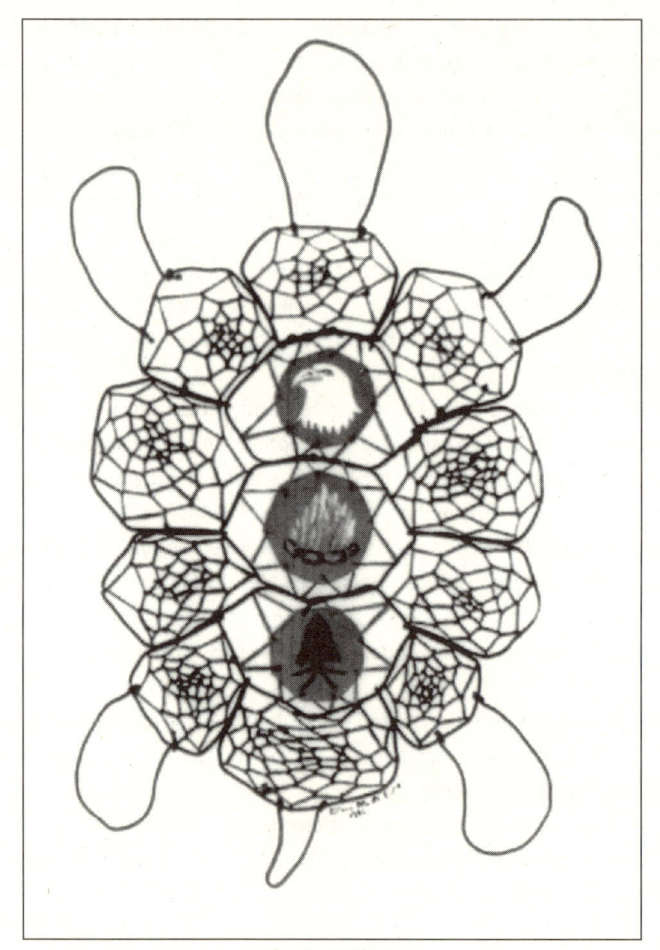

Turtle Island

3

Hoops of Spirituality in Science and Technology

NJOKI NATHANI WANE AND
BARBARA F. WATERFALL

When you educate a girl you educate a whole nation.
African proverb

Where I come from, God is a Woman; her name is Thinking ... She is called Grandmother ... and also Spider ... Where I come from, society is matrilineal, matrilocal, and matrifocal ... Where I come from, all spirituality is gender-based, and as near as historical, geological, paleontological, environmentalist, horticultural, or other measures show, the planet and the people are/were all the better for gender-based, elder-female-focused spiritual systems. No nuclear bombs, no toxic waste dumps, no chemical terrorism, no millions of lost species of our legacy. Warfare is actually prohibited, and pacifism, nurturing, healing, inclusiveness, egalitarianism among all members of the community of being, and profound spiritual and ritual awareness continue to characterize that system (Allen 1998, 89).

INTRODUCTION

We are women from two distinct Indigenous cultures, one Indigenous to North America, the land known as Turtle Island, and the other to Mother Africa. These backgrounds lead us to honour the teachings of our Indigenous traditions in our curricula and praxis. Prominent within Indigenous knowledge traditions is the inclusion of spirituality as a legitimate epistemological foundation. We view the bringing of spirituality to our research and educational practices as pivotal to a sustainable global future. We view our coming together as that of weaving and extending the knowledges that come from the traditions of our grandmothers. We extend this knowledge to each other and to you, the reader, across the expansiveness of two continents and from opposite ends of our Mother Planet's poles. In so doing we have enabled the unfolding of a circular ontology, that of a holistic understanding of being. This holistic understanding is inclusive of physical, mental, emotional, and spiritual realities. In our coming together we have reinforced our own Indigenous understandings with the awareness that all that exists around us is part of a great whole, namely that of the great hoop of Creation.

In this paper, we argue that spirituality is the science of our soul. We do not have a binary notion of science as distinct or separate from spirituality. Science as part of our Creation is infused with spirituality. The dimension and the magnitude of science is sought after and produced to satisfy the desire for Creation. The outcome of Creation is some form of technology. Within Indigenous societies, mothers and grandmothers create technologies to meet their everyday needs in the community. Technology thus conceived and executed cannot be separated from the spirit of its community. Our paper views Indigenous technologies as a force stemming from and functioning in harmony with the natural ecology. We thus argue that modern technology could be prudently applied to support and sustain both an ecological future, and the advancement of all peoples who inhabit the Earth.

We are quite aware that modern technology promotes fragmentation of relationships and creates independence and isolation. It also sustains colonialism and neo-colonialism through the opening of networks that were once restricted to individual communities; nations or continents have become globalized through the latest innovations in communication technologies. How then can spirituality be part of a commodity that promotes loss of identity, feelings of insecurity, and the entrenchment of inequality? In order to make sense of spirituality in a world of science, we employ a feminist, anti-colonial, discursive framework to expound the notion of using technology as if the world matters.

This paper reflects our personal transformative learning process of coming together as two Indigenous women from different cultures. By honouring and working within our own Indigenous cultural traditions, we were able to share our diverse perspectives. Rather than avoiding or fearing our diverse perspectives, we viewed them as a gift. By attending to ourselves as whole beings, we were able to engage with each other's power in a mutually respectful way. This proved to be a powerful and enriching experience. Most striking was that in sharing who we were as whole beings, we were able to re-affirm in a very real sense that, in spite of our diversity, we are all connected and related as a human family. We came to acknowledge what seemed like an old knowing – that we all come from the same source or root. We began to develop a deep spiritual sense that coming together and embracing our diversity is what we as human beings are being asked to do. It is thus the axiological standpoint from which our work came to be based. We invite you to join in on this circular understanding of being, as it will enrich further understanding. We do not propose to have all the answers; rather, our purpose is to rupture the hegemonic practice whereby technologies are predominantly used for destructive ends.

We contend that our Mother Planet is at a most critical juncture. We live within the context of a dominant colonizing capitalist and Eurocentric Western system. Within this context we, as members of the human family, are differently positioned subjects. We are positioned differently on the basis of race, culture and faith traditions, gender, sexual identity, class, and ability. As Smith (1987) and Ng (1999) argue, power relations are produced in everyday social interactions. Ng (1999) states that the exercise of power is manifested through examples of exclusion, marginalization, and silencing. From this perspective, we are all personally implicated whether our social interactions reproduce or actively rupture the dominant Eurocentric capitalist paradigm. We actively dismantle the dominant paradigm in our teaching by making use of transformative pedagogical approaches that embrace and honour the active practice of diversity.

To be fully human is what is required of us now on Earth. As Medicine Eagle states, "This means that more emphasis needs to be placed on being receptive rather than active; on relationship rather than on separation; on power in harmony with the great forces rather than focused on personal domination over others; in the darkness of winter's earthly germination as well in the rapid growth of summer days; on nurturing rather than fighting; on supporting rather than destroying; and on a deep and sacred ecology that deals respectfully and harmoniously with All Our Relations rather than with isolated issues" (1991, 117). We need to re-examine connections between the past, the present, and the future. We contend that this reconsideration needs to be grounded in a spiritual ontology. This inquiry can enable us to re-evaluate the hegemonic ways in which technological advancement has been employed and to seek alternative means based on Indigenous methods of creating life.

DISCURSIVE FRAMEWORK

Employing an Indigenous-centred, feminist framework, we evoke an anti-colonial discourse in order to theorize the inclusion of spirituality within science and technology. We conceptualize an anti-colonial, discursive framework as the absence of colonial imposition, as the agency to govern oneself, and the practice of such agency based on Indigenous foundational wisdoms (Dei 2000a, Fanon 1995, Trask 1991). Such a perspective ruptures the predominant Euro-Western scientific paradigm as the only legitimate axiological, ontological, and epistemological standpoint for research and technological development. In evoking an anti-colonial perspective we broaden our understanding of imperialism to include the cognitive. Smith (1999) articulated imperialism as that which "draws everything back to its centre" while distributing materials and ideas outward. Through this process, knowledges and cultures were to be discovered, extracted, appropriated, and distributed. Western cultural knowledges and science have been the "beneficiaries" of the colonial objective with respect to Indigenous peoples. Smith (1999) contends that the discoveries obtained from Indigenous subjects were commodified as property belonging to the West. She also asserts that not only did colonialism mean the imposition of Western rule over Indigenous lands, but it also meant the imposition of Western rule over all aspects of Indigenous knowledges, languages, and cultures.

The underlying assumption of our framework is that groups are located differently in the social hierarchy and that it is from these unique locations that each group produces knowledge about its understanding of the world. As subjugated and differentiated social groups, Indigenous peoples typically have been ignored or discredited both in academia and in mainstream society (Dei 2000b, Smith 1999, Ashcroft 1995). This paper intends to contribute to a growing body of literature that recognizes the connections between Indigenous knowledges, spirituality, science, and technology. We centre alternative modes of knowledge production that are rooted in the Earth-based framework of our grandmothers and mothers. This discourse emphasizes the foundational wisdoms generated by Indigenous cultures and traditions and uses these wisdoms to interrogate oppressive forces (Wane and Waterfall 2001, Smith 1999, Steady 1987).

We infuse an Indigenous feminist perspective into an anti-colonial discourse because of its holistic nature; its emphasis on the connectedness of the four elements of life that are wind, fire, water, and earth; and its focus on nurturance and life-sustaining principles. Central to most Indigenous feminist epistemologies is a consideration of diversities and differences inherent in emotional, physical, and social phenomena and an understanding of spirituality as a central and grounding force. As indicated earlier, we are two women who have learned the importance of inclusivity, the celebration of diversity, and the embracing of difference as fundamental both to meaningful connection and to the expansion of knowledge. We contend that the acknowledgement of our cultural differences and our unique colonial experiences contributes to the wealth of knowledge production in advancing elements of spirituality in science and technology. Through the process of meeting, we have awakened to a profound realization of our connectedness and our commonalities and to the understanding that our differences enhance our learning. We have also come to appreciate the differences and similarities in Indigenous ways of knowing.

DEFINING SPIRITUALITY

Our understanding of spirituality is rooted in the local ways of knowing of our communities. As such, it is informed by the teachings of our grandmothers and mothers. We could not therefore articulate notions of spirituality and technology without a discussion of Indigenous knowledges. (Please see Marie Battiste's more thorough discussion of Indigenous Knowledges in her chapter of this book.) The Indigenous knowledges valued in Aboriginal societies derive from multiple sources such as traditional teaching, empirical observation, and revelation. These categories overlap, interact, and enhance one another (Castellano 2000). Indigenous knowledges are also understood to be the "common sense ideas and cultural knowledge of local peoples concerning the everyday realities of living" (Dei 2000b, 2). Spirituality is embedded in the principles, facts, procedures, and systems of traditional knowledges of a given group of people. Spirituality is often culturally specific, meaning that its principles and practices were developed in response to the needs of a certain people in a specific environment. However, we have come to realize in our interaction as two women scholars from two distinct Indigenous communities that spiritual knowledges can be applicable to other cultures or people. In our dialogue, it became clear that in a world that is shrinking due to technological advancement, spirituality might be one of the few cultural threads able to maintain the interconnectedness of the Indigenous peoples of the world.

It needs to be stated that our understanding of spirituality is not representative of all Indigenous peoples from Turtle Island, nor of all Africans. By its very nature, spirituality is a personal enterprise. Our understanding of spirituality encompasses a holistic epistemological understanding inclusive of the mind, body, and spirit connection. Spirituality thus is not something separate from each other or from the world around us. Spirituality is about connection, relationship, belonging, and being as one within a universal system of kinship ties. Spirituality is all around us. Our spirituality is connected to our natural ecology, to our indigeneity, and to the community-based social actions emerging from where we are situated. We experience our spirituality as fundamentally experiential and intuitive rather than conceptual. As indicated by Hiatt (1986), spirituality comprises direct experiences of being rather than abstraction and reasoning, and is thus not an aspect of thought.

We take the position that *spiritual* is not a synonym for *religious* because a religion is an institutionalized set of beliefs, practices, and rituals regarding spiritual concerns and issues (Krippner and Welch 1993). One can be religious and spiritual or spiritual and not religious or religious and not spiritual. In this paper, we employ the term *spiritual* not from a religious perspective, but from a perspective that is Indigenous to us. Spirituality contains certain qualities of mind, including compassion, gratitude, awareness of a transcendent dimension, and an appreciation of life, all of which bring meaning and purpose to existence (Remen 1993, Vaughan 1991).

Spirituality is understood to be the determination that the mind produces, known as power of the will, to refine our behaviours through our own positive thoughts, words, and actions. We understand thus that spirituality creates positive productivity for self and others. Spirituality, by definition, is goodness, love, and compassion for all of humanity in both tragic and optimistic circumstances. For some people spirituality is intuitive and can be found in small acts of kindness, the beauty of flowers, the blue sky, rivers, or the singing of birds. Spirituality is our daily lifestyle and practice of truth, trust, love, and wisdom. It is woven through all life and is not just invoked occasionally. It characterizes the relationship of an individual to the universe and does not necessarily require a formal structure or ritual. Spirituality is found everywhere, not only in temples, churches, synagogues, mosques, stars, music, song, dance, nature, and intimate relationships, but also in every moment of everyday life (Taylor 2001). That is why we cannot separate spirituality from technology or science.

As Indigenous scholars teaching in the academy, who integrate the spirituality of our grandmothers and mothers in our daily practice, we acknowledge that speaking about this unspoken, unwritten spirituality in academic terms is, in a way, violating its fundamental epistemology. Yet we feel that this academic discussion is justified given its possibility for opening up new ways for others. When interviewed by Wane (2000), an Embu Elder in Kenya was willing to share:

Spirituality is not something out there, something that you read in your books, something that you can learn in your university. It is in your heart, it is in you. Spirituality is when you wake up in the morning and call on *Ngai* (Creator in Kiembu/Kikuyu) to guide you and to be with you. Spirituality is when you stand on your threshold and acknowledge the land you are standing on, the air you breathe, the creation surrounding you. Spirituality is when you give thanks before you eat or drink to Ngai first and to your ancestors, and to the land. In those brief moments when you pour a libation and welcome people to share what you have. For me that is spirituality.

We too understand our spirituality as a way of being, of connecting with the land, the universe, and creation. It is a state that is rooted in who we are as a people and manifested in our everyday activities and relations. This forms the discursive foundation of our spiritual rootedness. Our spiritual knowledge has been passed down both by word of mouth and through observation as a creative and dynamic force within us, around us, beneath us, and above us. To be spiritual thus is to maintain an awareness of this dynamism as it moves through and around our being. It is the practice of utilizing this creative force for the collective good of humanity.

A spiritual journey requires a commitment to transform one's consciousness from ignorance, denial, or grandiosity into harmonious alignment with one's divine nature. We begin by attending to ourselves as whole beings and move outward to the world around us. This understanding assumes that attending to spiritual dimensions contributes to positive personal growth and facilitates the harmonious bringing together of diverse perspectives. Our methodological practice begins with prayer, acknowledgement of our ancestors and the spirits of this land, and asking for the spirits' direction and intervention.

As Indigenous women we understand spirituality as an inner experience and thus perceive the fruits of spirituality as stemming from an epistemology of inner knowing. Our knowledges come from a very old and deep place, for they germinate from within the womb of creation – the place where all life begins (Waterfall 2002). Indigenous science and technology come from this space and are thus rooted in the centrality and sacredness of life itself. Grounded in a woman-centred ethic of nurturance and life giving, science and technology can develop to preserve and maintain a sustainable future for the seven generations to come.

THE POLITICS OF INCLUDING SPIRITUALITY IN SCIENCE EDUCATION

The culture, values, and traditions of [Indigenous] people amount to more than crafts and carvings. Their respect for the wisdom of their elders, their concept of family responsibilities extending beyond the nuclear family to embrace a whole village, their respect for the environment, their willingness to share all these values persist within their own culture even though they have been under unremitting pressure to abandon them. (Mr Justice Thomas Berger, Mackenzie Valley Pipe Inquiry, the Berger Inquiry, quoted by Doak 1997).

As we attempt to define spirituality, several issues emerge, such as those to do with culture, values, traditions, land, and relationships. In Wane's (2000) study of Embu rural women in Kenya, spirituality was often associated with the natural processes of life: the land, the universe, and creation. Thus, aligning spirituality with the values of a profit-driven society is not easy; in such a culture, human beings are spiritually disconnected from nature. O'Sullivan (1999) points to the massive destruction wrought by humans who conceive of themselves to be separate entities from the earth. Battiste and Henderson (2000) contend that this conceptualization of humans as separate from the natural environment is a core characteristic distinguishing Euro-Western from Indigenous scientific traditions.

Given the reality of Euro-Western scientific hegemony, it must be understood that the issue of centring Indigenous knowledges and perspectives inclusive of spiritual realities within science education is a politicized project. There is also an inherent danger of Indigenous expressions of spirituality being appropriated and commodified. Indeed, Smith (1999) speaks to the problematic of Indigenous knowledges becoming colonized and fragmented through processes of re-arranging, re-presenting, and re-distributing local knowledges for purposes that serve imperialistic ends. As indicated earlier, Indigenous knowledges function to sustain and support the natural ecology for the betterment of "All Our Relations." This is the axiological standpoint, or from our perspective, the *womb-base* from which Indigenous knowledges come into being. However, Euro-Western science and technology have been primarily used for a profit motive and inherently have a divide-and-rule component. Thus the outcomes of Euro-Western science and technology are not accessible to all people. Rather they are accessible only to those who can afford the accrued benefits of technological advancement. Therefore, a science curriculum that includes spirituality must speak both to inherent power politics and to inherent colonizing problematics.

We recognize that technological advancements such as the Internet are providing worldwide access to identical information. As cultural boundaries collapse, Western understandings and attitudes become dominant. While more and more people are concerned about the increased commodification of communication technology, the loss of identity, the feeling of insecurity, and the entrenchment of inequality, Indigenous peoples of the world are seldom considered in the discussion (Smith and Ward 2000). While advancement in technology has created unprecedented opportunities for some people, for Indigenous peoples, technological advancement threatens to exterminate and contaminate their traditional knowledges. In addition, this advancement threatens to extend the process of colonization, thus giving rise to the possibility of new forms of marginalization.

For many non-Indigenous people, technological advancement is a new way of opening markets for their goods and services and also for finding ways of commodifying and commercializing Indigenous knowledges, such as traditional healing practices and relaxation techniques such as yoga and meditation. In a sense, globalization has made available to a wider audience Indigenous cultures of the world. For Indigenous peoples, this means that they have to fight harder on a variety of fronts to ensure their cultural survival and to find new ways of asserting their rights and autonomy in the face of new threats posed by technological advancement (Smith and Ward 2000). It is imperative that Indigenous peoples establish linkages with other people who are Indigenous around the world. That is, Indigenous peoples must globally unite by identifying common goals arising from their lived colonial experiences and by creating a global arena for radical social and political change. We contend that given the present global economic context, this is an important way for Indigenous peoples to become active agents in the anti-colonial cause.

It is quite clear that the perception of Euro-Western science as the only way to understand experience, combined with Enlightenment humanism, has been largely responsible for the demise of the spiritual and moral dimensions of the curriculum. We contend that the centring of spirituality within science education serves a crucial purpose – to enable students to understand the domain of human experience that might differentiate human beings from the kinds of advanced intelligence machines that are likely to develop during the twenty-first century. We must not continue to allow scientific discoveries arrived at through technological development to proceed independently of a social context in which ethics and values play a defining role, as occurred in the twentieth century with the development of nuclear technology. In the twenty-first century, biotechnology is emerging as an issue of immense importance for science and ethics (Martin 2001, Hart 2001). We maintain that the purposes of science education should be to a) produce a body of scholarship through which knowledge and understanding of human spirituality and values might be advanced; b) create awareness of why spirituality is important in science education; and c) contribute to the on-going debate on how to save planet Earth (Martin 2001).

Because Euro-technology promotes fragmentation of relationships and creates independence and isolation, our challenge as scholars is to find a way to introduce spirituality into this faceless and feelingless, yet very powerful field. Because of technology, younger generations do not need to be taught by their grandmothers or Elders about their cultures. They can log into the Internet and find out for themselves. We question how the Elders will know that the teachings that these children are receiving are the right teachings. It is imperative that we find ways of rupturing the predominant Euro-Western paradigm. The centring of diverse understandings of spirituality, inclusive of Indigenous knowledges within the educational curricula can serve this vital function. We believe that as educators, we have a responsibility to encourage the development of the spiritual nature of students and to establish and enhance their awareness of their relationship with other species and with the land. Education approached in this way would broaden students' sense of community and help them maintain a respect for the centrality of life.

By engaging the spiritual dimensions of our students within the educational process, we open the door to valuing individual uniqueness regardless of race, sex, class, creed, or ability. Standard hegemonic education disregards the non-material, so that individuals are judged only on the basis of their measurable and quantifiable skills and talents. Those students who can follow the rules obediently are rewarded, whereas those who do not are paralyzed (Miller 1991) by being slotted into a system that disempowers them, perpetuates their inaccessibility to resources, and reinforces the relations of power and dominance held by certain groups. We contend that to effectively attenuate this problematic, teachers must understand the holistic nature of education, and the educational system must actively work to develop students' spirituality. In order to do this, there must be an acknowledgement that all students have spiritual potential. This, we contend, is a critical starting point for scientific education. By opening the door of understanding to the interconnectedness we have with each other, we come to science and knowledge production linked by this commonality, rather than separated by our differences. We contend that this educational approach will yield a more accepting and inclusive environment and way of thinking. It will impact on the practices of science by reconnecting people with holistic and life-sustaining ways of thinking, thereby creating sustainable development and earth-friendly technologies.

PEDAGOGICAL IMPLICATIONS

The project of constructing pedagogy that encourages spiritual growth within an academic setting presents itself with interesting challenges. Academia has not yet developed a structure that would support instructors who seek to provide elements of other forms of learning or knowledge. As Indigenous educators, our goal is to nurture the spirituality of our students and to respect their relationships with nature. Although spirituality is often not overtly spoken about in our classrooms, it is infused in everything we do – an emotional spirit characterizes the totality of the classroom process, from the very beginning to the end. This emotional spirit is experienced in every moment and in every interaction. Students at the beginning of class are usually asked to get in touch with themselves as whole beings. This can be done by narrating an incident or story that captures the students' emotions. Sometimes we tell stories of our personal educational experiences from our Indigenous perspectives. We tell of sitting around the fire in a circle, receiving teachings from our elders, grandmothers, and mothers. We also make students aware of the disruption of Indigenous education processes that has occurred through colonial imposition (Semali and Kincheloe 1999). This enables students to locate themselves in the educational politic and to connect themselves bodily, emotionally, and spiritually within the classroom.

Sometimes we bring to class inspirational readings or ask students to spend two or three minutes thinking about themselves, their relationships with other people, the universe, and their surroundings. At other times, students are asked to sit still and to meditate, or, if they cannot meditate, to think of nothing. Students may not understand why these pedagogical approaches have been brought into the classroom. Some students may interpret these practices as intellectually soft; attitudes of humility and admissions of ignorance are considered signs of weakness. However, others are able to connect emotionally and say that they use similar approaches in their teaching, or in their everyday activities. Classes can begin with a formal prayer, a cleansing ceremony of burning Indigenous medicines, or singing. The drum has also been brought into the classroom to remind us of our connection with the Earth and all that exists in the web of life. The vibration of the drum is itself a pedagogue. For instance, the sounds of the drum bring messages to us and we interpret these messages in our own unique way in accordance with our cultural understandings. Indeed, we contend that there is a transformation that takes place invisibly.

Informed by an anti-colonial discursive framework, our curriculum speaks to the problematic of colonialism and to the devastating effects that colonialism has had on Indigenous peoples and Indigenous communities. Many of our students' lives are wrought with pain and oppression. We have come to understand that we have a moral and an ethical responsibility to create an educational space that encourages the connection of mind, body, emotion, and spirit. As such, our pedagogical practices rupture the dominant Euro-Western academic paradigm, which attends only to the mind or intellect. Students are invited to come into the classroom as whole beings, where the body, emotion, spirit, as well as the intellect are acknowledged.

Reflective processes characterize our interaction with students in relation to the context and the content of the material we are dealing with. This approach enables each student to have an internal dialogue and in this process to develop a holistic relationship with him or herself. Our curricula and pedagogical practices also actively encourage students to personally reflect on where we are located in the dominant social hierarchy and to see ourselves as personally implicated in social relations of inequality. Informed by the wisdom of our ancestors, much of the classroom process utilizes the circle as the primary pedagogical tool. We have observed that when students are engaged in a holistic way within the context of the circle, a more meaningful dialogue takes place with other members of the classroom. Furthermore, dialogue within the context of the circle becomes more open and respectful, enabling racist, sexist, classist, elitist, and heterosexist social relations of inequality to be actively ruptured.

We recognize that a curriculum that speaks to the reality of colonial oppression and social relations of inequality can trigger the expression of deep emotion. Rather than discouraging or diminishing feelings that evoke tears, we encourage students to express their feelings within the context of classroom circles. Students learn valuable social skills as they are encouraged to support their peers through their expression of deep emotion in a safe and non-judgmental atmosphere. Through this process, a holistic educational experience becomes integrated into the classroom, providing rich soil for the development of nurturing, life-giving, and sustainable practices and technologies.

Indigenous peoples of the land, Indigenous methods of knowledge production, and Indigenous technologies are naturally infused with the ecology that is around them. Students who attend Euro-Western schools surrounded by concrete and treeless lots are at a disadvantage with respect to forming a connection between their spirituality and nature. Yet classroom pedagogical processes can easily link up with our ecology even on a small scale. Engaging in an indoor seed-planting project can cultivate an ethic of kinship responsibility and an understanding of reciprocity by allowing students to experience how seeds grow into plants which clean the air. The students in turn reciprocate by continuing to nurture the life of the plant. Connecting this exercise with expressions of gratitude to reinforce the spiritual connection can foster and strengthen a student's spirituality. On a larger scale, the class could embark on a tree-planting exercise on university property or in a neighbourhood park. This exercise would make a contribution to the university community, to the neighbourhood, and even to the local businesses, while helping students to develop a holistic sense of connectedness and relatedness with the world around them.

The holistic nature of the spiritual journey from within an Indigenous context often takes place in an outdoor environment. Nonetheless, through a series of inventive steps, a bridge between the outdoor environment and the classroom can be built. Two Trees (1993) brings natural artifacts that are imbued with meaning, such as branches and rocks, into the classroom. Next, she rearranges the physical space of the class into a more inviting and accommodating shape, such as a circle without the interference of chairs or tables. Finally, the entire classroom space becomes transformed when the students themselves engage in activities that explore their spirituality (Two Trees 1993).

We as Indigenous educators have employed similar pedagogical practices. Indeed, in order to maintain the context of Indigenous knowledges, Waterfall encourages students to bring in blankets and cushions so that they can sit comfortably on the floor. Wane encourages students to strengthen Indigenous perspectives by drumming or inviting guests from the community. Candlelight can also transform the classroom, enabling holism and creating a rich context for sharing and for the exchanging of knowledges. Furthermore, we as instructors have brought our sacred bundles or natural relatives into the classroom. In addition, Waterfall has organized field trips to various communities and wilderness sites in order to give students an opportunity to share and learn within nature and around a fire and to receive teachings from Indigenous elders and teachers.

CONCLUSION

The oral teaching of our mothers and grandmothers are the pillars of Indigenous women's spirituality. From the beginning, their spirituality has been intertwined with nature and its processes. This connection between spirituality and nature exists today among many Indigenous peoples of the world. Many Indigenous peoples have maintained their connection to the land, the universe, and creation through thought, words, and deed. Because of their intimate relationship with the land, they understand and respect the reciprocity between different life forms that are essential for maintaining the balance of life.

At this point in Earth's development, the issue of world peace is of urgent concern. We contend that an infusion of spirituality into science and technology can be used to help create a world of peace. World peace requires that we bring life-giving and life-sustaining spirituality to the forefront. It is imperative that we bring this understanding into everything that we do, into everything that we touch. In the contemporary academy, the cultivation of spirituality is largely absent. It is our conviction that Indigenous knowledges, sciences, and technologies must be brought to the centre at this time for the health of the planet, for a sustainable future, and for seven future generations to come.

REFERENCES

Allen, P.G. 1998. *Off the reservation: Reflections on boundary-busting, border-crossing loose canons.* Boston: Beacon Press.

Ashcroft, B.G. and Tiffin, H. 1995. *Post-colonial sudies reader.* London: Routledge.

Battiste, M. and Henderson, J.Y. 2000. *Protecting Indigenous knowledge and heritage: A global challenge.* Saskatoon, SK: Purich Publishing.

Castellano, M. B. 2000. Updating Aboriginal traditions of knowledge. In *Indigenous knowledges in global contexts: Multiple readings of our world*, eds. G.J.S. Dei, B. Hull, and D. R. Rosenberg. Toronto: University of Toronto Press.

Dei, G.J.S. 2000a. African development: The relevance and implication of Indigenousness, In *Indigenous knowledge in global context: Multiple readings of our world*, eds. G.J.S. Dei, B. Hall, and G. Rosenberg, 95–108. Toronto: University of Toronto Press.

– 2000b. Rethinking the role of Indigenous knowledges in the academy. *International Journal of Inclusive Education* 4(2), 111–32.

Doak, M. 1997. New religious movements. Retrieved November 29, 2001, on the World Wide Web. *http://www//religiousmovements.lib.virgina.edu/nrms/naspirit. html.*

Fanon, F. 1995. National culture. In *The post-colonial studies reader*, eds. B. Ashcroft, G. Griffiths, and H. Tiffin, 153–7. New York: Routledge.

Hart, J. 2001. Emerging face of spirituality in Minnesota healthcare. Retrieved on November 29, 2001 from World Wide Web. *http://www//innerresourceenhancement.com/emerg_face.htm.*

Hiatt, J.F. 1986. Spirituality, medicine, and healing. *Southern Medical Journal*, 79(6), 736–43.

Krippner, S. and Welch, P. 1993. *Spiritual dimensions of healing*. New York: Irvington.

Martin, A. 2001. Project for the scientific study of spirituality and value. Retrieved on November 29 from the World Wide Web http://www//mailto:martinashley@uwe.ac.uk.

Medicine Eagle, B. 1991. *Buffalo woman comes singing*. New York: Ballantine Books.

Miller, R. 1991. *New directions in education*. Brandon: Holistic Education Press.

Ng, R. 1999. *Toward an integrative approach to equity in education*, unpublished paper. Toronto: OISE/UT.

O'Sullivan, E. 1999. *Transformative learning: Educational vision for the 21st century*. Toronto: University of Toronto Press.

Remen, N. 1993. On defining spirit. *Noetic Sciences Review*, 27: 41.

Semali, L.M. and J.L. Kincheloe, eds. 1999. *What is Indigenous knowledge: Voices from the academy*. New York: Falmer Press.

Smith, C. and G. Ward, eds. 2000. *Indigenous cultures in an interconnected world*. Vancouver: UBC Press.

Smith, D. 1987. *The everyday world as problematic: A feminist sociology*. Toronto: University of Toronto Press.

Smith, L. Tuhiwai 1999. *Decolonizing methodologies: Research and Indigenous peoples*. New York: Zed Books.

Steady, F.C. 1987. African Feminism: A Worldwide Perspective. In *Women in Africa and the African diaspora*, eds. R.Terborg-Penn, S. Harley, and A. Benton Rushing, 3–24. Washington, DC: Howard University Press.

Taylor, B. 2001. Seven principles of spirituality in the workplace. Retrieved on November 29 2001 from World Wide Web. http://www.itstime.com/rainbow.htm.

Trask, H.K. 1991. Natives and anthropologists: The colonial sruggle. *The Contemporary Pacific 3* (1), 159–67.

Two Trees, K.S. 1993. Mixed blood, new voices, In *Spirit, space and survival: African American women in (White) academe*, eds. J. James and R. Farmer, 13–22. New York: Routledge.

Vaughan, F. 1991. Spiritual issues in psychotherapy. *Journal of Transpersonal Psychology 23*(2), 105–19.

Wane, N. and B. Waterfall. 2001. Embracing diversity: A transformative learning exercise. Paper presented at the 4th Transformative International Education conference. Ontario Insititute for Studies in Education, University of Toronto.

Wane, N. 2000. Indigenous knowledge: Lessons from the elders: A Kenyan case study. In *Indigenous knowledges in global contexts: Multiple readings of our world*, eds. G.J.S. Dei, B. Hull, and D.R. Rosenberg, 95–108. Toronto: University of Toronto Press.

Waterfall, B. 2002. Feminist reflections on Wombma-based spirituality, In *Back to the drawing board: African Canadian women and feminism*, eds. Wane, N. et. al. Toronto: Sumach Press.

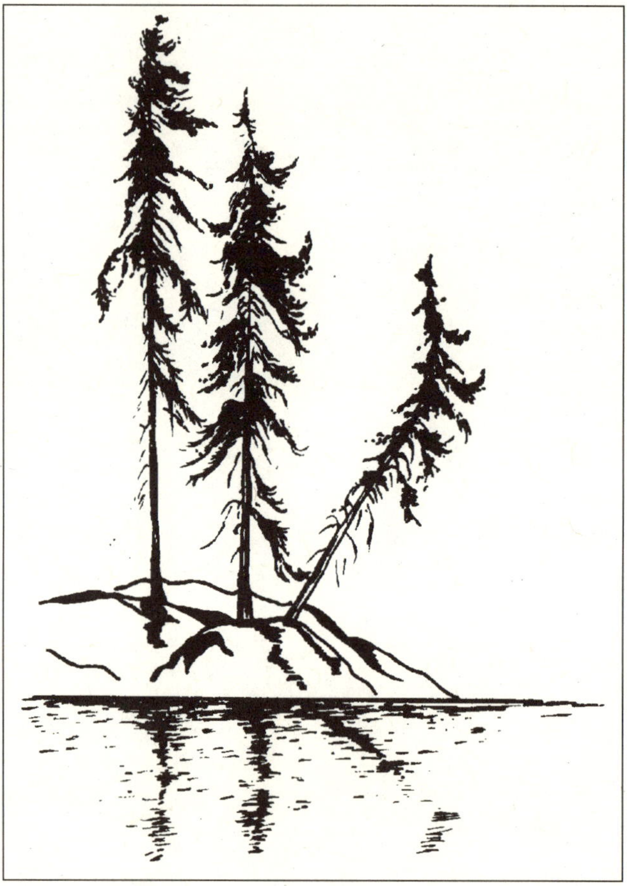

4

Teaching Sustainable Science

PEGGY TRIPP

My epistemological crisis as a science instructor started on my first day as professor at this northern university. The provocation was the continuous stream of lumber trucks that rolled by laden with softwoods destined for the mills that provide such an important economic foundation for this city and country. I learned by experience that the lumber trucks are continuous. They feed the mills that operate 24 hours a day, 365 days a year. But for me they created an emotional angst and confusion undefinable at the time. Every truck symbolized unsustainability on wheels to me, yet my profession and the entire economic system of the Western world were complicit. When I think back several decades to that first truck I watched with a sense of foreboding, I can still feel those unsettling emotions, now coloured by a greater understanding but also heightened by an ever-increasing sense of urgency. Added to my distress of witnessing the millions of trees that have passed by my window in that time is the professional knowledge of the ecosystem change in our boreal forest towards a less diverse, hardwood-dominated disturbed forest. These combined sentiments have altered my academic direction towards teaching and producing knowledge as if the world mattered.

In the first decade of my scientific work I was able to put aside my distress as irrelevant and bask in the success that comes from practising the traditional view of science esteemed for its independence from biases and values. With generous funding for forest research, I initiated and directed an active forest genetics research program, trained graduate students and post-doctoral fellows, and published. However, after a decade of successful research, seeds of doubt about the philosophical underpinnings of my own research program began to germinate. Funding became increasingly tied to industrial applications and the inadequacy of environmental management of toxic wastes from my laboratory became a more personal ethical dilemma. And have I mentioned the trucks? A sabbatical of reflection and reading opened a new world of ideas in the area of feminist critique of science. Suddenly I was able to access a vocabulary to understand and express my concerns about my own research. This work culminated in an analysis of gender bias in my scientific field of expertise (Tripp-Knowles 1994) and the decision to close my research laboratory (*Asking Different Questions* 1996) and redirect my scholarly work.

SCIENCE AS A CONTRIBUTOR TO AN UNSUSTAINABLE ENVIRONMENT

This change in professional direction gave me the unusual opportunity to explore science as an endeavour – an opportunity unavailable to or unsolicited by most scientists. Looking to the works of scientists was not fruitful for accessing an analysis or critique of science as a whole. As John Ralston Saul has noted, "The few scientists who learn to doubt and dare to do so in public are usually discredited as being unstable by the majority of the scientific community" (Saul 1992, 304). Rather, the areas of study that most resonated with me were those in the realm of philosophy and social science, those of critique spanning the topics of scientific objectivity, reductionism, methodology, the social and political construction of science (LaTour 1993; Longino 1990) imperialism, commoditization (Levins and Lewontin 1985), domination of nature (Leiss 1994) and the ethics/science relationship (or lack thereof). Historical influences of interest included the work of Thomas Huxley (defender of Darwinism) who predated the present refusal to recognize values expressed in science. He managed to infuse science with a 'virtuous' ethic for the purpose of transforming nature from its 'depravity' (Huxley 1959). This represented the scientific mood of the era – a yearning for control, not sustainability. More recently in the 1960s the relatively new biological science, ecology, was greeted with high expectations that it might provide a solution for sustainability. However, it fragmented in its development into a diversity of subfields with a commitment to "monitor" environmental problems but without a mandate to find agreement on solutions (Worster 1994). An example, adding texture to the exploration, of the impact of science on sustainability comes from the geographer, John Whitelegg, who showed that the scientific reasoning applied to the problem of saving time in human transportation systems invariably increased pollution (Whitelegg 1995). My overall conclusion from my study was that science – as a method, as the backbone of technology, as a social and political construct, and particularly as the popularly worshipped saviour for societal ills – is a significant contributor to environmental degradation, and thus a detriment to sustainability.

Such a conclusion leaves a professional scientist with a conundrum, particularly for teaching. How could I continue teaching science to students and contribute to a sustainable world at the same time? The answer to this dilemma took me back to Thomas Huxley's penchant for considering the intersection of morality and science. (As an aside, my admiration of Thomas Huxley has expressed itself in my life by naming a feline family member in his honour). Perhaps the goal of transforming science by incorporating a "care ethic" in place of his "subduing nature ethic" would translate into the classroom. It was a tall order but worth a try and the first step was to look to the literature for a theoretical foundation and potential models of a caring or sustainable science.

First, I pursued a definition of sustainability among the innumerable uses and abuses of this term. The most widely used definition comes from the World Commission on Environment and Development that states "sustainable development is development that meets the needs of the present without compromising the ability of future generations to meet their own needs" (1987, 43). I prefer David Orr's emphasis away from development towards a sustainable continuing relationship between humans and nature in his comprehensive description of ecological sustainability (1992). His definition includes the concepts of involving citizens in creating a continuing future, widespread ecological literacy, recovering subjugated Aboriginal knowledges, using nature as a model for design of human constructions, reconsidering issues of scale and centralization, and recognizing the human/nature interrelatedness. My use of the term *sustainability* incorporates these holistic concepts. Thus sustainable science refers to those components that relate to science as activism, particularly knowledge production about nature.

THEORETICAL PROPOSALS ABOUT SUSTAINABLE SCIENCE

I discovered that one of the origins of the term "sustainable science" is an outgrowth of feminist science scholarship. The connotation of social and environmental responsibility in this phrase is deliberate. Donna

Haraway's concept of "situated knowledges" was an attempt to define a knowledge from a feminist perspective; that is, as a "situated and embodied knowledge" as opposed to "various forms of unlocatable, and so irresponsible knowledge claims" (Haraway 1996, 255). She describes her vision of a feminist science that has an intentional undertone of ethics and politics rather than an emphasis on epistemology: "Feminists have stakes in a successor science project that offers a more adequate, richer, better account of a world, in order to live in it well and in critical reflexive relation to our own as well as others' practices of domination, and the unequal parts of privilege and oppression that make up all positions" (Haraway 1996, 252).

Another feminist scientist, Sandra Harding, develops a concept that she calls "strong objectivity," which also contributes to the discussion of social and environmental responsibility in science (1991). Objectivity is considered "strong" when there is a commitment to examining all the values and politics impinging on a scientific research project. This includes the consideration of consequences of the work. Moreover, this analysis must be an integral part of the process of knowledge production and not just an assessment after project completion. As Harding herself describes: "[T]he sciences need to legitimate *within scientific research* (author's italics), as part of practicing science, critical examination of historical values and interests that may be so shared within the scientific community, so invested in by the very constitution of this or that field of study, that they will not show up as a cultural bias between experimenters or between research communities" (1991, 146–7). Such an analysis as part of the scientific endeavour would shift the direction of science away from studying *how we know* towards the study of *what it is that we know* as well as *why it is that we know it*. This shift would be a step towards producing knowledge with the consideration of its social and environmental context. Such knowledge would serve to remedy what Sandra Harding feels is a conspicuous failure of science, that is, to show a "commitment to and [be] reasonably successful at increasing human welfare" (1993a, 49). This suggestion of incorporating social responsibility in scientific endeavours is a notable contribution. However, Harding's concern is restricted to humans and is thus problematic.

A third feminist theorist, Carolyn Merchant, prioritizes the relationship of humans to the natural world by developing the notion of the "partnership ethic" that is grounded by the concept of relation (1996). She proposes that understanding nature is best done when the human community and the biotic community are in a mutual relationship with each other. Social and environmental responsibility is incorporated into this idea because "the greatest good for the human and non-human community is to be found in their mutual living interdependence." (Merchant 1996, 216).

These three perspectives on attending to responsibility in knowledge production are examples of "sustainable science." Londa Schiebinger (1997) cites this term "sustainable science" as originating from Helmut Hirsch and Helen Ghiradella (1994) who used it in a Native American science journal to express the idea of an endeavour that not only meets the needs of the present human and natural community but that also can be sustained for many generations. Another common element in the above arguments of Haraway, Harding, and Merchant is the use of gender as an analytical category. Thus, Londa Schiebinger considers that a robust feminist science could and should be a sustainable science, a science that goes well beyond the limitations of women's equality to include the many ideals of various feminisms such as environmentalism, humanism, and pacifism (1997).

Two other views relevant to a sustainable science come from the environmental ethics domain of Eurocentric philosophy as well as Indigenous knowledges. Cheney and Weston (1999) question the very basis of Western ethics as it is presently conceived. They argue that Western ethics is epistemology-based ethics that sees the world as composed of facts to which we respond with a system of ethics. It is the goal of science, of course, to discover these 'facts.' Cheney and Weston, however, propose that the world we inhabit actually emerges from our ethical practice. From this angle, what we know is really dependent on how we treat the 'object' of our inquiry. Ethics are therefore foregrounded as fundamental to our understanding and knowledge production. In other words, ethics comes first, and then knowledge. They propose that we incorporate "etiquette" (or caring) as the foundation of our knowledge production. Knowledge thus becomes intimacy and reciprocity. They call this proposal an ethics-based epistemology founded on etiquette. A consequence of this alternative perspective is that love is paradigmatic, that is, love is in fact a way of knowing. Similarly, trust is a crucial component to knowing. This concept of etiquette as a foundation of knowledge production complements the proposals of feminist scientists with their proposals of caring and social responsibility as fundamental to a sustainable science.

Finally, Indigenous knowledges provide models for a sustainable science. Here, Michael Wassegijig Price, a member of the Wikwemikong

First Nations, contemplates knowledge production and teaching science from his experience:

> The foundation of Western science is measurement, using numbers and empirical equations to describe and predict natural phenomena. At one time in the distant past, the goal of Western science was seeking universal truths. But, today, perverted by the industrial revolution and capitalism, the goal of Western science is about what sells in the marketplace. Indigenous knowledge is another way of understanding reality. Indigenous knowledge is not based on measurement or quantification but rather on relationship and observation. Unlike Western science and technology ... [Indigenous knowledge] has tenets of sacredness and spirituality. These ideas directly affect our relationship to and interaction with nature and one another. Thus we are not just invisible, objective observers but actual and accountable participants in the complex web of life ... This interaction will maintain and protect the long coexistence between the Anishinaabe and the natural world (Price 2000,18–9).

This participatory way of knowing, which incorporates respect and caring, is indeed a sustainable way of knowing and knowledge production.

COURSE DEVELOPMENT FOR TEACHING SUSTAINABLE SCIENCE

This richness of theory and models for sustainable science from the literature provided me with confidence, and even a licence, from my perspective, to redesign my science teaching to incorporate notions of sustainability. However, tackling the science curriculum in higher education was another matter. An easy approach would have been the equivalent of liberal feminism, that is, "just add women and stir." In other words, I could just add a segment or assignment about sustainability to a traditional science course without altering the foundational assumptions of the course materials. This retained too much of a traditional mode of knowledge construction to appeal to me. The approach I preferred would comprise a completely revised alternative course founded squarely on sustainability in science. Such a course would be unlikely to receive the approval of colleagues with more traditional disciplinary backgrounds. This dilemma was not easily resolved. Rather, the course resided on my professional "wish list" in the back of my mind. In hindsight, this waiting approach was the right solution, because it allowed the opportunistic presentation of the idea of nontraditional courses at those rare but valuable departmental discussions of curriculum revisions. I will briefly describe three of the circumstances that allowed me to design and offer courses that are non-traditional in my academic context.

The first course is a critical analysis of science as an endeavour offered to higher level (usually fourth-year) students in a biology degree program. This emphasis "about science" – rather than "doing science" – makes it non-traditional and thus low in priority for a department stretched for faculty resources. However, the retirement of a colleague who had instructed a course on how to read and understand the classic primary literature in biology provided an opportunity. The course was called Science under Scrutiny and would be deleted from departmental offerings if no one assumed its instruction. I was pleased to oblige this request and immediately and completely revised the content to encompass the interdisciplinary science critique literature. This new content was predictably not highly valued by my departmental colleagues. Thus, my motion in a subsequent departmental meeting to revise the calendar description to better reflect the new content of my course was defeated. By this time, however, the popularity and positive reputation of the course among the students ensured its retention with high enrolment despite the misleading calendar description.

The second opportunity arose when faculty resources were stretched in another department with which I am associated. Along with most other faculty, I was asked to teach a course well beyond my area of expertise. This in itself was not a serious problem for me since the road to my doctoral degree taught me well how to learn. However, the course had an applied component to it that had a direct link with what I considered to be unsustainable natural resource management. I took a risk by refusing to teach the course and as a result was invited to resign my partial appointment with that department. Fortunately this closing door became an opportunity to pursue an appointment with another department, Women's Studies, where I was pleased to develop two courses, Ecofeminism and Feminist Science, both of which incorporate social responsibility and sustainability as foundational components.

The third opportunity to teach a non-traditional course arose with subsequent negotiations to repair my association with the department from which I had resigned. Due to economic cutbacks, faculty resources were continuing to diminish and so my assistance was requested. The new department head asked for my input on course offerings. I proposed a course about environmental ethics that would simultaneously be offered to two student groups, one in the present department emphasizing resource management and another in a faculty emphasizing resource preservation (or "tree cutters" and "tree huggers" respectively). My suggestion met with the department head's approval. An informal verification of faculty response was undertaken by the department head. He reported back to me that the faculty in general did not support such a course. I was surprised that he was surprised at this finding. He proposed that if I was willing to take another risk, I could prepare such a course to present to the faculty for their approval. I felt that the risk of preparing a course without a definite opportunity for use was well worth taking at this particular moment because I knew that cross-listed courses would meet the administrative demand for fiscal restraint by increasing class sizes. By a slim margin, the course was approved with a title that did not use the words "environmental ethics." Not surprisingly, in a matter of a few years, trends have shifted so greatly that an environmental ethics course with the actual title of Environmental Ethics is now required for many students in the resource management program.

In summary, I would underscore opportunism and its corollary, patience, as valuable strategies for effecting curriculum change. Of equal importance I would propose two other tactics, those of embracing interdisciplinarity for course content and of celebrating marginality as a professional strategy in an academic setting. Further, I propose the following five strategies for teaching sustainability in the higher education classroom.

STRATEGIES FOR TEACHING SUSTAINABILITY

Making visible that science is not value-free

The most important guideline that I use for teaching science students about sustainability and its relationship to science is also the most difficult for students to grasp. It is analogous to teaching fish about water. That is, science is embedded in values that are generally invisible to many students, like water is to fish. In other words, there is a political agenda in every scientific undertaking that is important to recognize and evaluate. Harding's notion of "strong objectivity" makes this very point. Without a realization of the points of conflict between science and sustainability, there cannot be an appreciation of a mode of knowledge production that deliberately incorporates sustainability. In my teaching about feminist science, we, as a class, 'discover' that science has a history of bias that is sexist, classist, racist, and speciesist. This particular course emphasizes the social responsibility component of sustainability by addressing the first three of these four oppressions. My course about ecofeminism stresses the fourth, discrimination against non-human species. In my Science under Scrutiny course for science majors, the students have the disadvantage of having been immersed in science studies for so many years that the 'value-free' pretense of the discipline has taken on near-impermeable dimensions. Thus we dwell on uncovering the values and assumptions that underlie the scientific enterprise that they have been studying. The significant take-home message is that science is not value-free as it is popularly touted, and that science and technology may conflict with sustainability. This message does not carry with it solutions for what a sustainable science would look like. However, the awareness of the interaction between science and sustainability is a crucial first step towards envisaging a new philosophy of science that incorporates "caring" or "etiquette" as proposed by theoreticians.

Centring learning in students' lives

Educators concerned with environmental issues, critical thinking, and sustainability have noted the importance of connecting education directly to individual student's lives. For example, Freire proposes that "epistemologically it is possible, by listening to students speak about their understanding of their world, to go with them towards the direction of a *critical* (my italics), scientific understanding of their world" (Shor and Freire 1987, 106). A teaching tool intended to link sustainability to students' lives is the use of personal journals. The 'text' for each of my courses comprises a reading packet of scholarly papers for each class discussion. In preparation for each class, students are asked to read each paper and respond in a variety of ways, one of which is to write a two-page journal entry about reactions to their reading. They are encouraged to consider connections between the reading and their personal experiences. We discuss levels of analysis of personal experience with the goal of reaching beyond superficial interpretations. To the extent that the papers under consideration deal with some aspect of sustainability (as many do), the students are reflecting on issues linking sustainability to their own lives.

Another requirement of most of my courses that grounds learning in students' personal experiences is the student presentation. The topics are chosen by students themselves (from within a relatively broad domain) with the requirement that part of the presentation must include an analysis of why that topic is meaningful to the student. Further, for my courses pertaining to environmental ethics, students are required to consider the framework of ethical stances presented in lecture and reading format to determine which ones coincide with their personal value system and why. This provides them with a vocabulary for contemplation and discussion. (In many cases the discussion precedes the contemplation!) This is a particularly heuristic strategy in a course that includes students from programs with different environmental values. Student evaluations make it clear that one of the most valuable components of this course is the airing and comparing of personal values about the environment. Of particular interest to students is the delicate discussion about which values, if any, are 'better' and why. This addresses the question posed by the environmental educator, David Orr, "Should we strive to teach values appropriate to sustainability, or should we present these as only one possible orientation to the world?" (Orr 1992, 142). A final teaching strategy that is grounded in students' lives is discussion about personal biocentric life styles. One of the liveliest discussions occurs around the question, "How can we as instructors and students leave a small footprint on this world?"

Proposing power redistribution

Moving from the personal to the global, I make the point that teaching and learning about sustainability requires an appreciation of the history and hegemony of national and international power structures and ideologies. marino (1997) defined hegemony in this context as the process of entrenching social relations that advantage powerful elites in such a way that very few people stop to question the situation, thus acting in complicity with it. Such entrenchment directly affects the environment and thus sustainability. Bowers (1996) delineates a variety of hegemonic assumptions including the privileging of rationality, consumerism, an ideology of progress, colonialism, worship of science and technology, individualism, and the faith in limitless natural resources. Incorporating such material is a daunting task in courses that are not primarily focused on politics and economics. I like to consider the topic of colonialism as a doorway to the exploration of power redistribution, because its relationship to social and environmental abuse is easy to grasp. From the environmental perspective, a paper by Callicott (1989) about Indian land ethics along with its critiques (Hester et al. 2000) is a good point of departure. A publication by Battiste and Henderson (2000) underscores the critical importance of respect and protection of both Indigenous knowledge and heritage. The consideration of colonialism from the scientific perspective is well documented by the books, *The 'Racial' Economy of Science* (1993b) and *Is Science Multicultural?* (1998) by Sandra Harding (edited and authored respectively). And from the ecofeminist perspective, examples of discussions of colonialism include the many works by Vandana Shiva (e.g., 1989) and Taylor's writing about environmental justice and ecofeminism (1997). In short, I insist that a sustainable future depends on the recognition of the ubiquitous nature of colonialism that forms such an integral part of our Western, Eurocentric world view. In my teaching experience, discussions among students of these issues culminate in a commitment to and respect for cultural diversity along with a resistance to the many other expressions of global power imbalances.

Hope and empowerment

Consideration of sustainability in the classroom inevitably raises emotions such as frustration about environmental degradation. Addressing students' sense of despair and powerlessness in the classroom was never a consideration in my teaching agenda. It was not a focus of my education, nor are discussions about such issues prominent in the scholarly literature with which I was familiar. Only when comments about hopelessness escalated to the point of obviating student engagement in course material was I motivated to address the issue directly. As professors are wont to do, I turned to the literature for exploration and set aside classroom time for consideration of the issues with students. One discovery was the work of Ann Fausto-Sterling (1986), who noted an improved learning environment in her science classroom as a result of discussing emotional reactions to content material. Study materials suitable for fostering hope are not numerous but are available. The best of these is a qualitative study of 100 people who have had long careers in the paid workforce pursuing issues involving social and/or environmental responsibility (Daloz et al., 1996). Notably these individuals were chosen for their continued enthusiasm and optimism and the study explores their ability to maintain this enthusiasm. Other works include Bernard and Young's *Ecology of Hope* (1997) that describes activists working towards sustainable communities. David Suzuki's book entitled *Good News* is just that, and a striking departure from his many previous books with their cautionary message. Another work with a simple but profound message is by the Dalai Lama (Gyatso 1992). In my experience, this type of reading and discussion material improves the students' learning (not to mention their professor's learning). A final strategy for empowering students is a course component on individual action toward some sustainable community enterprise. Students submit a report describing goals and accomplishments for community work (usually unpaid) that they undertake during the semester on behalf of a sustainable world. It is intended that this commitment to action, pursuit of a meaningful task, and analysis of the value of the work is part and parcel of empowerment, assisting both self-understanding and understanding of community organizations.

Learning about anti-speciesism or intrinsic value

This final strategy for encouraging students to consider the significance of sustainability in the classroom and beyond consists of allocating class time to the topic of addressing discrimination against non-humans. This concept, called "anti-speciesism" within the sociological framework and "intrinsic value" within the area of environmental ethics, is not a new one for most students. What is novel is the allocation of class time to addressing this issue and the many connections that can be made to other well-learned concepts. For example, students familiar with the disciplines of women's studies and sociology value the connection of anti-speciesism to the other forms of inclusiveness, anti-sexism, anti-racism, and anti-classism. Ecofeminist readings (e.g., Merchant 1996) and discussions about what is a care ethic are meaningful to these students. Natural resource management students are excited by the connection of anti-speciesism to organisms beyond animals. They respond well to reading and discussions based on the question posed by Stone (1974) in the title of his paper, "Should trees have standing? Toward legal rights for natural objects." And science students ponder this issue in the context of scientific objectivity. "Is there or should there be a sanctity in nature that has been removed by the scientific enterprise?" Readings such as Merchant's *The Death of Nature* (1980) and discussions about ways of knowing in other cultures (Siu 1957; Nasr 1968), including Aboriginal ways of knowing, support these discussions. This leads to discussions about transforming science or creating a new way of perceiving the role of science for a sustainable future. The theoretical proposals described above inform such discussions. From my perspective, these are the pivotal readings and discussions for teaching and learning about sustainability. Until we incorporate anti-speciesism into our psyche and discourses by appreciating the intrinsic value of non-humans, sustainability will elude us as a society. From my perspective, this issue is too important to ignore in our daily lives and in the microcosm of our world, the classroom.

The trucks are still rolling by my window as they did over two decades ago. The harvested trees making up their load are now noticeably smaller than they used to be, from more northern distances. Ordinarily this would be reason for increased anxiety but my angst has diminished over time. I am inspired by the energy of my students and their commitment to sustainability. I also try to follow the advice that I give to them: Take time to pursue hope with reading and discussion and become empowered with community activist work of your choice for sustainability. And finally, don't look to the magnitude of change in the system for rewards from your efforts. Rather, reward is in the creation of opportunities for expressing your values and in the satisfaction of expressing them well. It is a privilege to have the opportunity to teach and learn about sustainability in the classroom.

REFERENCES

Asking Different Questions: Women and Science. 1996. Directed by Gwynne Basen and Erna Buffie. 57 min. Artemis Films/National Film Board of Canada. Videocassette.

Battiste, M. and J.Y. Henderson. 2000. *Protecting Indigenous knowledge and heritage: A global challenge.* Saskatoon, SK: Purich Publishing

Bernard, T. and J. Young. 1997. *The ecology of hope.* Gabriola Island, BC.: New Society.

Bowers, C.A. 1996. The cultural dimensions of ecological literacy. *Journal of Environmental Education* 27(2),5–10.

Callicott, J.B. 1989. American Indian land wisdom? *Journal of Forest History*, Jan., 35–42.

Cheney, J. and A. Weston. 1999. Environmental ethics as environmental etiquette: Toward an ethics-based epistemology. *Environmental Ethics* 21, 115–34.

Daloz, L.A.P., C.H. Keen, J.P. Keen and S.D. Parks. 1996. *Common fire: Leading lives of commitment in a complex world.* Boston: Beacon Press.

Fausto-Sterling, A. 1986. Women and minorities in science: An interdisciplinary course. *Radical Teacher* 30, 16–20.

Gyatso, T. 1992. A Tibetan Buddhist perspective on spirit in nature. In *Spirit and nature,* eds. S.C. Rockefeller and J.C. Elder, 109–24. Boston: Beacon Press.

Haraway, D. 1996. Situated knowledges: The science question in feminism and the privilege of partial perspective. In *Feminism and science,* eds. E.F. Keller and H.E. Longino, 249–263. Oxford: Oxford University Press.

Harding, S. 1991. *Whose science? Whose knowledge: Thinking from women's lives.* Buckingham: Open University Press.

Harding, S. 1993a. Forum: feminism and science. *National Women's Studies Association Journal* 5(1), 47–55.

Harding, S. (ed.) 1993b. *The "racial" economy of science.* Bloomington: Indiana University Press.

Harding, S. 1998. *Is science multicultural?* Bloomington: Indiana University Press.

Hester, L., D. McPherson, A. Booth and J. Cheney. 2000. Indigenous worlds and Callicott's land ethic. *Environmental Ethics* 22(3), 273–90.

Hirsch, H. and H. Ghiradella. 1994. Educators look at contemporary science teaching. *Winds of Change.* Spring, 38–42.

Huxley, T. 1959. *Man's place in nature.* Ann Arbor: University of Michigan Press (reprinting of original, 1863)

Klein, N. 2000. *No logo.* Toronto: Vintage Canada.

Latour, B. 1993. *We have never been modern.* Cambridge: Harvard University Press.

Leiss, W. 1994. *The Domination of nature.* Montreal and Kingston: McGill-Queen's University Press.

Longino, H. 1990. *Science as social knowledge.* Princeton: Princeton University Press.

Levins, R. and R. Lewontin. 1985. The commoditization of science. In *The dialectical biologist,* eds. R. Levins and R. Lewontin, 197–208. Cambridge, MA: Harvard University Press.

marino, d. 1997. *Wild garden: Art, education and the culture of resistance.* Toronto: Between the Lines.

Merchant, C. 1980. *The death of nature.* San Francisco: Harper and Row.

Merchant, C. 1996. *Earthcare: Women and the environment.* New York: Routledge.

Nasr, S.H. 1968. *Science and civilization in Islam.* Cambridge, MA: Harvard University Press.

Orr, D.W. 1992. *Ecological literacy.* Albany: State University of New York Press.

Price, Michael Wassegijig. 2000. Of science and spirit. *Tribal College* 12(2):18–21

Saul, J.R. 1992. *Voltaire's bastards.* Toronto: Penguin Books.

Schiebinger, L. 1997. Creating sustainable science. *Osiris* 12, 201–16.

Shor, I. and P. Freire. 1987. *A Pedagogy for liberation: Dialogues on transforming education.* New York: Bergin and Garvey.

Shiva, V. 1989. *Staying alive.* London: Zed Books.

Siu, R.G.H. 1957. *The tao of science: An essay on western knowledge and eastern wisdom.* New York: MIT Press and John Wiley and Sons.

Stone, C.D. 1974. *Should trees have standing? Toward legal rights for natural objects.* Los Altos, CA: Kaufmann, Inc.

Suzuki, D. and H. Dresser. 2002. *Good news for a change.* Toronto: Stoddart.

Taylor, D. (1997) Women of color: Environmental justice and ecofeminism. In *Ecofeminism: women, culture, nature,* ed.K Warren, 38–81. Bloomington: Indiana University Press.

Tripp-Knowles, P. 1994. Androcentric bias in science: An exploration of the discipline of forest genetics. *Women's Studies International Forum* 17(1), 1–8.

Worster, D. 1994. *Nature's economy.* Cambridge: Cambridge University Press.

Whitelegg, J. 1995. The pollution of time. In *Science for the earth: Can science make the world a better place?* eds. T. Wakeford and M. Walters, 115–46. New York: John Wiley and Sons.

World Commission on Environment and Development. 1987. *Our common future (Report of the Brundtland Commission).* New York: Oxford University Press.

5

Professional Ideology and Educational Practice: Learning to Be a Health Professional

MOIRA M. GRANT

NOW YOU SEE ME ...

I appear in this discussion as a medical laboratory technologist who is researching my profession as part of my doctoral research. I reflect on how I have arrived here and on my own struggles as I attempt to draw on new perspectives to understand my profession as part of a larger scientific (and imperfect) discipline, and to consider possible directions for its educational processes.

I first encountered medical laboratory work as a summer student in a downtown Toronto hospital and, after obtaining my professional certification, I continued to practise in other hospital laboratories in the province of Ontario. I worked with instrumentation that seemed complex, yet beautifully, logically linear. I relished the independence with which I conducted my analyses, often as the sole worker on night shifts in the laboratory. Later, I taught in a program of medical laboratory technology and parented my children while doing consulting work in health sciences education. I look back on that period and see my unceasing efforts to keep learning – there wasn't a time that I wasn't registered in at least one academic course. And I wrote, constantly. What had been a mere dalliance with playful fiction in high school turned into a passionate commitment to making my own learning processes accessible to others. Through my writing, *I became visible*. The issue of visibility was to become an important theme in my research.

I turned to further degrees in education. All the while I was aware of a distinct unease – a sense of separation from medical laboratory technology as I explored various aspects of adult and health professional education. I felt I had to make a choice between education and medical laboratory technology. I experienced an initial resistance to opening myself to critical perspectives on my own profession. I struggled with the question of where to situate myself – an insider who definitely felt like an outsider to my own profession – as I asked questions about medical laboratory technology. I was touched by the work of Freire (1995) but felt vaguely uncomfortable envisioning my somewhat privileged colleagues as the "oppressed." However, it was the work of Ann Bishop (1994) that drew me into the notion of working against oppression as an ally from within my own profession. My head swam with delight and awe at the discovery of the immense bodies of literature, such as the feminist critiques of science and Third World feminism. This is where I see myself as I approach this consideration of medical laboratory technology.

I find evidence of my transformation in the development of my own thesis research. A project that started out as a purely document-based historical analysis of medical laboratory technology became a dialogic process as I began to seek perspectives from my medical laboratory colleagues on their own profession and, later, as I reflected those perspectives back to them through presentations and publications. My initial conceptualization of a dichotomy in my own career – medical laboratory technology *or* education – has melded into what seems to me to be a very powerful partnership of two passions. As a result, I understand the dichotomization of the world from a very personal perspective, and I am excited to find it contributing to my understanding of medical laboratory technology even as it still insists on playing hide-and-seek with my growing awareness of the world.

MAKING MEDICAL LABORATORY TECHNOLOGY VISIBLE

Medical laboratory technologists perform analyses on body fluids and tissues to provide information for diagnosis, treatment, and monitoring of disease. They comprise the third largest health profession in Canada, and yet they play a virtually invisible role in the health care system. There are over 21,000 medical laboratory technologists in Canada, approximately 87 per cent of whom are women (Canadian Society for Medical Laboratory Science [CSMLS], 1999; CSMLS, 2001) and most of whom are White Western Europeans (Grant, 2004). Yet, despite the huge contribution to health care decisions that can be attributed to the work of medical laboratory workers (Department of Health, 2002; Kratz and Laposata, 2002; Richardson, 1999), the profession is better appreciated through the concepts of devaluing, distancing, dehumanizing, and disconnection.

In his study of medical laboratory technology in eastern Canada in the late 1800s, Twohig (1999) used a framework of equality and collaboration to describe the emergence of this occupational group as part of the 'health care team.' However, a more critical analysis of this female-dominated profession suggests this framework is unhelpful. McKenzie (1992) has maintained that women were preferred for the job of laboratory testing in the late 1800s and early 1900s because of their assumed patience for detailed work, superior personal hygiene, and ability to concentrate during the painstaking testing process. Other probable conditions were women's submissiveness to male authority and their willingness to tolerate poor salaries (less than what their male counterparts earned) and minimal opportunities for career advancement (Kotlarz, 1998a). Victorian notions of cleanliness and racial purity (Kraut, 1994) resulted in an almost complete exclusion of immigrants from consideration for these jobs. In the early 1900s, the laboratory became one of the few sites where women could work in the science field because opportunities for women and racialized groups to work in medicine and science were eliminated by the closure of some of the specialized university medical programs in Canada and the U.S. These closures have been attributed to Flexner's (1910) critique of medical education, and resulted in reduced educational opportunities in science and medicine for women (Brouwer, 1999; Schiebinger, 1999), and Catholics, Jews, and Blacks (Cassedy, 1991), among other groups. Shepherd (1993) suggests that the very science that marginalized groups were called upon to facilitate in the early days of scientific practice was actually being used to disenfranchise them through sociobiological arguments, psychological theory, and so-called 'scientific' theories about appropriate social roles for women and immigrants.

Although medical laboratory technology was promoted in its early days as a collaborative venture with physicians (perhaps as a strategy to attract women to the profession), doctors kept a tight rein on laboratory workers, initiating all analyses, and supervising and approving their work at every step (Kotlarz, 1998b). The work of laboratory technicians, as these workers were called before the establishment of formal professional education and credentialling, was seen as requiring little thought. Many outsiders felt that there was no point in formalizing technicians' education, believing that a brief training period with the hospital pathologist (a physician who specializes in the identification of disease) would suffice to provide the necessary mechanical skills.

And so it began ... white women in a profession cobbled together to perform routine testing procedures with which physicians no longer wished to tarry as more sophisticated and glamorous techniques beckoned, women whose work was highly structured and demarcated by the needs and values of the medical profession, and whose competence as scientists was never even considered a possibility. Medical laboratory technology has remained under the watchful gaze of the medical profession from its very inception. Even with the introduction of formal medical laboratory educational programs in Canada and the United States in the 1920s and 1930s, the curricula and accreditation processes were – and continue to be – highly influenced by the medical profession as well as by the dictates of the scientific method. Most licensing requirements for laboratory departments still require that laboratories be overseen by a pathologist. The research tradition among medical laboratory technologists in Canada is weak, and the technologists who *do* publish research do so often as second or third authors in publications oriented to the medical profession or in publications outside their own profession. Research is highly quantitative and follows the traditional IMRAD (Introduction, Materials and Methods, Results and Discussion) format that has arisen from the scientific method. The stance of the impartial and distant observer is very much a part of the intra-professional dialogue and evidences much of the detachment, objectivity, rationality, and reductionism so strongly critiqued in the feminist science literature (for example, Benston, 1982). In particular, the abstractness of mathematics is ubiquitous, not only in the numerical format of most laboratory reports, but also in the statistics used in research publications and quality assurance techniques, evidence of what Shulman (2001) refers to as the "mathematico-technological" characteristics of Western culture (415).

The medical laboratory profession operates within the shadow of medicine. All the efforts and products of its practitioners' work are shielded from the public, resulting in limited public awareness of what they do. When laboratory testing is ordered by a physician or designate, the results are reported to and interpreted by the physician for the patient. The role of the technologist is rendered invisible. For example, in laboratory practice, the patient is usually introduced to the technologist as an accession number, a bar code, or one of a batch of specimens. This distancing and dehumanization of their work means that technologists see their purpose as providing accurate and timely results to within plus or minus two standard deviations of the mean – the only outcomes they are permitted to see – rather than as having an impact on the health and well-being of the patient. And it is not only the technologists' considerations of patients that are dehumanized – in a recent discussion of laboratory costs, laboratory technologists themselves were objectified through their placement in the section on *"issues* relating to the quality of laboratory performance" (Richardson, 1999, 69) (emphasis added) rather than in the section on laboratory personnel. Medical laboratory technology suffers from a marginalization that is seen in its physical isolation from patients – usually a remote laboratory removed from patient activities – and from the daily operations of health care institutions (exacerbated by the privatization and outsourcing of laboratory services), and in the business and statistics-oriented language of laboratory operations. This dehumanizing trend is also evident in Canadian health care restructuring over the last two decades: laboratory workers have been reduced to the status of 'FTES' (Full-time equivalents) for staff scheduling purposes and shuffled from one workplace to another in an effort to rationalize laboratory operations and maximize efficiency.

It is not just the medical profession that exerts a powerful influence on medical laboratory technology; the business world has also shaped laboratory practice. While medicine has contributed evidence-based medicine – which has been criticized for its reduction of patient care to rules, probabilities, and statistics in order to rationalize medical decision-making through reliance on scientific evidence – business models exacerbate a dehumanized and distant stance with such icons as quality-of-life indicators (Siegrist and Junge, 1989), Total Quality Management (Armstrong and Armstrong, 1996), accountability (White and Begun, 1996), patient outcomes (Smith et al., 1996) and efficiency and cost-cutting in health care (Burke and Stevenson, 1998; Stein, 2001; White, 1997). Schiebinger (1999) comments on this grim focus on efficiency that seems to accompany scientific practice. She sees laboratories as data factories and describes a 'necropolis ideal' in which ideas are kept entombed and unexposed in order to maintain perfection and control. The overall effect is to reduce professional practice to statistics, with patients as variables and technologists as the means to improved outcome measures. Recent market-oriented shifts in laboratory operations to for-profit models have only served to exacerbate the perception of medical laboratory technologists as little more than expense figures in a budget.

All the while that technologists toil to assume the veil of objectivity through their labours, the very nature of their work as inherently *manual* is discounted and discountable. Hubbard (2001) notes the historical valuing of mental labour over manual labour, writing that "the people who work with their hands (as well as, often, their imaginations) are the ones who perform the operations and make the observations that generate new hypotheses and that permit hunches, ideas, and hypotheses to become facts" (154). Schiebinger (1999) acknowledges the tendency in science to devalue physical work when compared with intellectual work, and it is this privileging of intellect that is part of Benston's (1982) "myth of objectivity." In fact, Benston notes that the isolation of rationality leads to the objectification of human beings and, as I see it, to the belief that manual work *cannot* co-exist with intellectual work in the same person.

It would seem, then, that the very same system that defines medical laboratory technology also confines it. Indeed, medical laboratory technologists have historically seen quite a sharp distinction drawn between their practice (technology) and what physicians, microbiologists and chemists do (science). My efforts to encourage technologists to write about their work experiences and observations often meet with the response "But I'm just a lab tech." Even medical laboratory technologists themselves do not see the intellectual investment they make in the work they do.

Since the early days of incorporation of the national medical laboratory professional societies in both the United States and Canada, there has been an ongoing discussion about what medical laboratory technologists can rightly call themselves. In Canada, use of the word "medical" in the professional title was hotly contested by physicians in the 1930s, and even now laboratory practitioners in both countries debate the issue of who has the right to call themselves "scientists." Significantly, many of those who draw the most exclusive boundaries on science, claiming that medical laboratory technologists are *not* scientists, are those who are firmly entrenched in its upper levels – physicians, microbiologists, and chemists.

Medical laboratory technology is seen by its leaders and practitioners as the ideal site for professional re-invention through adoption of and specialization in molecular genetics techniques. However, such a move only reinforces the existing reductionist tendencies of the profession through the fragmenting re-organization of knowledge that Spanier (2001) sees accompanying an ideology founded at a molecular basis. Weasel (2001) has suggested that molecular biology does not reveal its biases readily because of the small scale on which it operates. She maintains that the distancing that inevitably follows breaking life down into increasingly smaller components renders impossible an appreciation of the living organism as a whole. Weasel questions prevalent scientific practices such as immunology and cell biology that use militaristic and mechanistic metaphors of control, domination, separation, and disconnection. She critiques as a "severe form of reductionism" (427) the assumption that it is necessary to kill something in order to understand it.

This notion of disconnection and death is implicit in the practice of medical laboratory technology; laboratory techniques make use of a process of active experimentation. Lederman and Bartsch (2001) describe such activities as inherently androcentric in their manipulation of a situation in order to test a hypothesis, and differentiate them from 'passive' experimentation, whereby an individual observes a natural situation. Medical laboratory technology involves almost exclusively *in vitro* ("in glass") techniques that necessitate removing the material to be tested from the patient's body. If the material is blood or tissue, the cells in these specimens begin to die the moment they are removed from the patient's body, echoing Weasel's comments about knowledge through death. Furthermore, such specimens are extracted from the patient in painful, possibly intrusive, and often embarrassing procedures conducted in sterile, dehumanized surroundings, a reality critiqued strongly by Haraway (1997). These practices are normalized and unquestioned in the unassailable pursuit of scientific knowledge as an ultimate good.

Benston (1982) has suggested that the problem with science is not that it contains bias, but that it makes the assertion that there is no bias. Certainly, medical laboratory practice embodies that assertion in its almost exclusive implementation of the scientific method. Kerr (2001) has acknowledged the scientific elitism that accompanies such objectivity-seeking practices, maintaining that not only do they obscure questions about who benefits, but that they also safeguard the status quo and existing power structures. The same objectivity that privileges validation by a professional community in order to ensure security (manifested in the laboratory as accuracy and reproducibility) also creates isolation and distance.

SEEING A NEW SCIENCE IN HEALTH PROFESSIONAL EDUCATION

According to Eisenhart and Finkel (2001), science reform has addressed mainly the most liberal of feminist critiques through compensatory strategies to lower barriers to the sciences, but without addressing the more radical critiques about the questions that scientists pose, the presence of bias in the foundations of science, and the privileging of one perspective over others. Kerr (2001) cautions against assuming that the apparent perspicacity of feminist critiques, especially with respect to methodology, has influenced research practice. She cites the isolation and small numbers of feminist researchers in the sciences as challenges to educational change. She suggests that feminist epistemologies may be problematic to women in some science disciplines (notably the natural sciences), particularly with their approaches to reflectivity and objectivity. The idea that knowledge is socially constructed may fit only with difficulty into the natural sciences.

The following discussion is not intended to be an exhaustive presentation of ways in which health professional education might benefit from new perspectives on science, but simply offers a few examples of the general directions it might begin to take. Given that professions such as medical laboratory technology have not even begun to acknowledge or discuss the feminist critique of science, some of these suggestions may represent a greater challenge to educators in this profession than in others.

Focusing on conventional science for transforming the curriculum is not likely to bring about the kinds of change that feminist critics of science are advocating (Eisenhart and Finkel, 2001). It is rather unlikely that substantial reform can be achieved across professions by introducing feminist critiques only from within the traditional science curriculum or after it has been developed. But since educators will probably not spontaneously embrace feminist critiques and methodologies, there may be more potential for change in an approach that first introduces dialogue about *how* science is practised in the profession and about assumptions that have traditionally gone unquestioned. Encouraging students and practitioners to question their own practices may prompt an eventual reconsideration of the relevance of educational curricula.

For example, tolerance for ambiguity is rarely found in scientific practices. Yet it is possible to encourage in students an appreciation that there may be many valid alternatives to a problem rather than one 'correct' solution, even in a subject area such as mathematics that might once have been seen as tolerating few grey areas (Shulman, 2001). Shulman advocates the use of open-ended problems for students, while Schiebinger (1999) suggests promoting an environment where "both/and" thinking rather than "either/or" thinking is nurtured. When I first came across the suggestion that there could be "grey areas" in mathematics, my first reaction was to balk at the inexactness of such a concept. I questioned how reasonable it would be for a medical laboratory technologist to issue a laboratory result that read "this analyte *could* have a value of 5.0 units, but then again it could also be 4.0 or 6.0." It was then that I realized that *this is exactly what a laboratory result does say*, but the inexactness is hidden in the tacit statistics of laboratory reporting – where a test result might indeed be recorded as 5.0 units, *people in the know* realize that the result is not meant to be interpreted as exactly 5.0, but as most likely falling within a statistical range of probability above and below the reported number. Even the supposedly exact practice of statistics has an implicit ambiguity that is not acknowledged openly by its practitioners. To point this kind of assumption out to students and non-scientists, and in so doing to teach them, as Shulman suggests, to *expect* standpoints and ambiguity in scientific statements, could help demystify science and open the door to a greater tolerance of questioning and grey areas.

Kerr (2001) advocates a tolerance of diversity in research methodologies and maintains that the methods of the social sciences cannot simply be imported into the natural sciences as a means of broadening scientific practices. Interdisciplinary research, and an openness to research approaches that are novel, at least for the laboratory profession, can introduce a breadth of experience and perspective that will strengthen, rather than bring into question, the validity and objectivity of a research project. My observations of research in medical laboratory technology affirm that the field is almost exclusively quantitative; this is certainly an area where educational interdisciplinarity could serve to introduce openness to, and indeed expectations of, a variety of research methodologies and epistemologies. Implicit in this new approach to teaching research would be exposure to and examination of the power relations that divide practitioners of medical laboratory technology from the 'experts' (who are usually not medical technologists). This is dangerous territory, to be sure, as it challenges the privileged positions of scientific experts, but I believe that challenging science's race- and gender-blind 'master narratives' (Barton, 1998) is absolutely essential to creating a new and living laboratory science.

Weasel (2001) has proposed introducing new metaphors into science education that stress connection, cooperation, and interrelationship. For example, she suggests redefining the cell membrane as a semi-permeable and dynamic zone of interaction between the cell and its exterior, rather than as a barrier to hostile influences or to loss of self. She notes that many states of disease, and cancer is a good example, occur when cell communities are deprived of their interconnections with the cell systems within which they originated. Weasel suggests that students would benefit from being guided to an appreciation of the value of interrelationships, cellular and otherwise, in a holistic, systems-oriented curriculum. Introducing such perspectives could be valuable in creating a science that is less fragmented and more open to previously marginalized holistic world views (Battiste, 1998; Hampton, 1993; Sparrow, 2001).

This concept of interconnectedness would serve to combat the pervasive fragmentation of many current curriculum models in medical laboratory science. Education and practice in this profession have traditionally been divided into five main subject areas: chemistry, hematology, microbiology, histology, and transfusion medicine (once known as "blood banking"). These sub-disciplines are based on archaic and arbitrary criteria that, in some past time, may have seemed logical, convenient, or efficient. Now they simply maintain artificial boundaries and prevent a holistic appreciation of the body. And sad to say, students who benefit from curricula that take a more integrated approach to laboratory medicine are often confronted, upon graduation, with workplaces that are still organized according to the traditional five sub-disciplines.

Fragmentation in medical laboratory education is not confined to the subject matter. For example, it is not until the end of most programs that students have an opportunity to work in a real laboratory setting (the practicum). Until that point, students are still very much isolated from the health care environment and patients. The didactic curriculum reproduces the academic setting to which most students are accustomed from their high school and university experiences: lectures, seminars that often end up being lectures, and mock laboratories. Thus it happens that individuals who have chosen medical laboratory technology as their specialty may not experience a real laboratory setting until two or three years after they have made their pre-admission one-day visit to a hospital. The workplace setting is radically different from what they will have experienced in their mock laboratories. They will suddenly be confronted with situations that they heard about way back in first year. Without realizing it, they have been socialized into the role of student, and know nothing of what it is to be a medical laboratory technologist. They are confronted with a disjuncture between the didactic phase and the practicum experience that is enabled by a lack of congruence between the two learning settings. The divide is so marked that bewildered students have been known to confess that, once they get into the clinical setting they are told, directly or indirectly, "Well that was the way you learned it. Now we'll show you the way you're supposed to *do* it here."

Students entering their practicum may also unknowingly encounter a mismatch between the equity/power axis in their educational program and that in their practicum setting. The majority of teaching staff in medical laboratory technology programs is White and female, reflecting fairly accurately the 87 per cent female population of the profession. I believe in the capacity for female instructors to serve as role models for women in science. However, the administrative structure of health care organizations is markedly different in equity and power distribution from what is found in educational programs. Once they arrive in their work settings, students are more likely to see mostly White female technologists in positions subordinate to White male physicians, managers, and administrators; they will also see racialized groups, male and female, in subordinated positions in the health care hierarchy.

It is essential for educators to address issues of power imbalance within the health care system, and to help students to create their own, intimate relationships with science (Barton, 1998). The ahistorical and hyper-rational nature of science curricula, and the distancing that results, needs to be addressed before scientific disciplines can be truly open to the perspectives of women and other disenfranchised groups (Fausto-Sterling, 1991).

Feminist critiques are only beginning to address the issue of developing curricular strategies that can counter the fragmenting and distancing effects of traditional approaches to science education (Hughes 2000, Ladyshewsky and Edwards 1999, Rosser 1997, Titchen and Higgs 1999, among others). In the case of medical laboratory technology, an integrative approach will need to address, not just the subject matter of the programs, but the ways in which educational practices embody and perpetuate the rational, objectifying approaches of old ways of doing science.

... AND NOW YOU SEE MEDICAL LABORATORY TECHNOLOGY

In my first explorations of feminist critiques of science, I was influenced by Ursula Franklin's story of Dr Alice Hamilton, one of the first women physicians in the United States (Asking Different Questions, 1996). Dr Hamilton went to Chicago to work with patients of the typhoid epidemic in the late 1800s and began asking why so many poor families were getting sick. Her work with public health issues suffered the same marginalization that much of the holistic socially and environmentally oriented public health of that time experienced in a world of reductionist curative scientific medicine. But unlike her male colleagues, who were asking questions like "Is the child sick?" Dr Hamilton asked *"Why is the child sick?"* Simply by framing her question differently, she was able to obtain answers that helped her trace the source of the bacteria to broken sewer lines in the poorer areas of town that were receiving little or no maintenance or repair. Franklin suggests that those who are deeply immersed in traditional science settings, including medicine, often have no sense of the limitations of their practice, and as a result, are not aware of the wealth of wisdom that lies within reach if only they were to ask different questions from those delineated within their own disciplinary canons.

Medical laboratory technology, deeply embedded in medical and business models of science, remains invisible both to the health professions community and to its own practitioners, who have yet to shift the questions they ask from "What do we need to know to be good medical laboratory technologists?" to *"Why* do we need to know this?" and *"How* does this relate to the world as a whole?" The many possible answers promise to bring the profession into science's field of vision.

REFERENCES

Armstrong, P. and Armstrong, H. 1996. *Wasting away: The undermining of Canadian health care.* Toronto: Oxford University Press.

Asking different questions: Women and science. 1996. G. Basen and E. Buffie (Directors) Montreal, Quebec: Artemis Films in co-production with the National Film Board of Canada. [Videotape].

Barton, A. C. 1998. *Feminist science education.* New York: Teachers College Press.

Battiste, M. 1998. Enabling the autumn seed: Toward a decolonized approach to aboriginal knowledge, language and education. *Canadian Journal of Native Education,* 22(1),16–26.

Benston, M. 1982. Feminism and the critique of scientific method. In *Feminism in Canada: From pressure to politics,* eds. A.R. Miles and G. Finn, 47–66. Montreal: Black Rose Books.

Bishop, A. 1994. *Becoming an ally: Breaking the cycle of oppression.* Halifax, Nova Scotia: Fernwood Publishing.

Brouwer, R. C. 1999. Beyond 'women's work for women': Dr Florence Murray and the practice and teaching of Western medicine in Korea, 1921–1942. In *Challenging professions: Historical and contemporary perspectives on women's professional work*, eds. E. Smyth, S. Acker, P. Bourne and A. Prentice, 65–95. Toronto: University of Toronto Press.

Burke, M. and Stevenson, H. M. 1998. Fiscal crisis and restructuring in medicare: The politics of health in Canada. In *Health and Canadian society: Sociological Perspectives*, eds. D. Coburn, C. D'Arcy and G. Torrance. Toronto: University of Toronto Press, 597–618.

Cassedy, J. H. 1991. *Medicine in America: A short history*. Baltimore:Johns Hopkins Press.

CSMLS (Canadian Society for Medical Laboratory Science) 1999. Membership survey report – Part 1. *Canadian Journal of Medical Laboratory Science*, 61(1), 11–14.

— 2001. *Medical laboratory technologists national human resources review: A call for action*. Hamilton: Canadian Society for Medical Laboratory Science.

Department of Health. 2002. *Pathology – The essential service: Draft guidance on modernising pathology services. A consultation document*. London UK: Department of Health.

Eisenhart, M.A. and Finkel, E. 2001. Women (still) need not apply. In *The gender and science reader*, eds. M. Lederman and I. Bartsch, 13–23. London: Routledge.

Fausto-Sterling, A. 1991. Race, gender, and science. *Transformations* 2, 4–12.

Flexner, A. 1910. *Medical education in the United States and Canada: A report to the Carnegie Foundation for the advancement of teaching*. New York: Carnegie Foundation for the Advancement of Teaching.

Freire, P. 1995. *Pedagogy of the oppressed*. New York: Continuum Publishing Co.

Grant, M. 2004. Under the microscope: 'Race,' gender and class in Canadian medical laboratory science. Unpublished PhD thesis. Toronto: University of Toronto, Dept. of Theory & Policy Studies in Education.

Hampton, E. 1993. Toward a redefinition of American Indian/Alaska native education. *Canadian Journal of Native Education*, 20(2), 261–309.

Haraway, D.J. 1997. *Modest_witness@second_millennium: FemaleMan©_meets_ oncomouse™*. London: Routledge.

Hubbard, R. 2001. Science, facts, and feminism. In *Women, science, and technology*, eds. M. Wyer, M. Barbercheck, D. Giesman, H. Örun Öztürk, and M. Wayne, 153–60. New York: Routledge.

Hughes, G. 2000. Marginalization of socioscientific material in science-technology-society science curricula: Some implications for gender inclusivity and curriculum reform. *Journal of Research in Science Teaching*, 37(5), 426–40.

Kerr, E.A. 2001. Toward a feminist natural science: Linking theory and practice. In *The gender and science reader*, eds. M. Lederman and I. Bartsch, 386–406. London: Routledge.

Kotlarz, V. R. 1998a. Tracing our roots: early clinical laboratory scientists and their work – myth and reality. *Clinical Laboratory Science*, 11(4), 209–213.

– 1998b. Tracing our roots: Origins of clinical laboratory science. *Clinical Laboratory Science, 11*(1), 5–7.

Kratz, A., and Laposata, M. 2002. Enhanced clinical consulting – moving toward the core competencies of laboratory professionals. *Clinical Chimica Acta, 319*(2), 117–25.

Kraut, A. M. 1994. *Silent travelers: Germs, genes and 'the immigrant menace.'* New York: BasicBooks.

Ladyshewsky, R. and Edwards, H. 1999. Integrating clinical and academic aspects of curricula. In *Educating beginning practitioners: Challenges for health professional education,* eds. J. Higgs and H. Edwards, 88–93. Oxford: Butterworth Heinemann.

Lederman, M. and Bartsch, I. 2001. Creating androcentric science. In *The gender and science reader,* eds. M. Lederman and I. Bartsch, 63–7. London: Routledge.

McKenzie, S. B. 1992. History of clinical laboratory science education. *Clinical Laboratory Science, 5*(4), 221–6.

Richardson, H. 1999. Laboratory medicine in Ontario: Its downsizing and the consequences on quality. *Clinical Chimica Acta 290,* 57–72.

Rosser, S. V. 1997. *Re-engineering female friendly science.* New York: Teachers College Press.

Schiebinger, L. 1999. *Has feminism changed science?* Cambridge MA: Harvard University Press.

Shepherd, L.J. 1993. *Lifting the veil: The feminine face of science.* Boston: Shambhala Publications, Inc.

Shulman, B.J. 2001. Implications of feminist critiques of science for the teaching of mathematics and science. In *The gender and science reader,* eds. M. Lederman and I. Bartsch, 407–22. London: Routledge.

Siegrist, J. and Junge, A. 1989. Conceptual and methodological problems in research on the quality of life in clinical medicine. *Social Science and Medicine, 29*(3), 463–468.

Smith, J.A., Kilgore, M.L., Mennemeyer, S.T., and Steindel, S.J. 1996. *What patients really want: Quality of laboratory services and improvement of patient outcomes in an era of managed care. Abstracts from the 1996 CLMA laboratory-related measures of patient outcomes symposium.* Clinical Laboratory Management Association. Available: http://www.clma.org/research/96symabs.htm [1999, January 16].

Spanier, B. 2001. Foundations for a 'new biology,' proposed in Molecular Cell Biology. In *The gender and science reader,* eds. M. Lederman and I. Bartsch, 272–88. London: Routledge.

Sparrow, L. 2001. *Interest of Aboriginal students in postsecondary education.* Unpublished master's research project. Toronto: Ontario Institute for Studies in Education of the University of Toronto, Dept. of Theory & Policy Studies.

Titchen, A. and Higgs, J. 1999. Facilitating the development of knowledge. In *Educating beginning practitioners: Challenges for health professional education,* eds. J. Higgs and H. Edwards, 180–8. Oxford: Butterworth Heinemann.

Twohig, P. L. 1999. *Organizing the bench: Medical laboratory workers in the Maritimes, 1900–1950.* Ph.D. thesis. Dalhousie University.

Weasel, L. 2001. The cell in relation: An ecofeminist revision of cell and molecular biology. In *The gender and science reader,* eds. M. Lederman and I. Bartsch, 423–36. London: Routledge.

White, D.B. 1997. *The key to OD in health and human service organizations: Service Quality.* A paper prepared for the 27th Annual Information Exchange and the 17th Organization Development World Congress, July 15–19, 1997, at the University of Colima, Colima, Mexico.

White, K.R. and Begun, J.W. 1996. Profession building in the new health care system. *Nursing Administration Quarterly,* 20(3), 79–85.

6

Mainstreaming Transformative Teaching

ANN MATTHEWS

INTRODUCTION

Reflecting on my experiences teaching business courses to all-female classes at an Ontario Community College during the 1980s made me realize that the pedagogy I used in the classroom (curriculum content and instructional practices) reproduced the dominant power relationships that marginalize women in education and in work. In fact, my pedagogical practices put these students in the ironic position of having to accept the very systems that marginalized them. The professional development available during my teaching years did little to encourage a critical interrogation of classroom pedagogy. In subsequent years, the graduate courses I attended at OISE/UT (Ontario Institute for Studies in Education of the University of Toronto) defined the pedagogy I was using as a traditional pedagogy and introduced me to the notion of "transformative teaching"; a pedagogical approach that includes practices from feminist, critical, and anti-racist pedagogies. Although these pedagogies "have different roots and traditions, they do share a common goal: to expose existing inequalities and instill critical consciousness in students" (Ng 1998, 1). Transformative teaching would have encouraged my students to question rather than accept the dominant ideologies and power relationships that shaped their lives.

I regard transformative teaching as a form of activism in teaching. Teachers who engage in this method of teaching recognize that traditional pedagogical practices support dominant power relationships.

Traditional pedagogy maintains the existing patterns of stratification and differentiation in society, which marginalize students on the basis of gender, race, class, disability, sexual orientation, and other difference. Activism implies a challenge to this status quo. Transformative teaching is teaching to bring about social change. It is teaching that addresses issues of social justice by challenging so-called common-sense knowledge and assumptions. To conceptualize how I could begin to move from traditional pedagogical approaches to transformative teaching, I explored the literature on the subject. This paper synthesizes discussion from several sources to outline the theoretical underpinnings of traditional and transformative pedagogies and some of the pedagogical practices associated with each.

There is no cookbook solution for transformative teaching (Tisdell 2001). It is an ongoing learning process for both teacher and student. Varying interrelationships shape the outcome of the transformative teaching effort. Not every pedagogical practice is applicable in all classroom situations. The teacher's choice of pedagogical practices, for any particular class, is influenced by the context within and outside the classroom, the positionality[1] of the students and teacher within this context, and the course content. To more fully understand the ways in which transformative pedagogical practices impact the classroom and facilitate or impede change, teachers should familiarize themselves with debates that address both sides of the issue. Attention should be paid to discussions that suggest pedagogical practices that might be used to initiate change and also to discussions that question the limitations of these practices to do so (see Ellsworth 1992, Muzzin and Labreche 1999). Transformative learning is not a comfortable experience for either the student or the teacher. Inherent in its processes is a challenge to what is accepted as the norm. For this reason, teachers and students must be prepared to tread on new ground and engage in difficult self-reflective processes in transformative classrooms.

UNDERSTANDING PEDAGOGY

Pedagogy is most frequently associated with traditional instructional practices that give primacy to the teacher as expert knower and view students as passive recipients of knowledge. The focus, in this context, is on *what* is being taught rather than on the processes involved in learning and in the construction of knowledge. For this discussion, I move beyond this limited concept of pedagogy by using the definition developed by Kenway and Modra (1992). They describe pedagogy as "what is taught, how it is taught and how it is learned ... [Pedagogy encompasses] the nature of knowledge and learning...[including how] knowledge is produced, negotiated, transformed, and realized in the interaction between the teacher, the learner, and the knowledge itself" (140). Transformative teaching embraces Kenway and Modra's definition, by focusing on all pedagogical processes and the interconnections between them. As teachers we need to create opportunities through which transformation may happen. In this discussion, I focus on opportunities for transformation that may be created through instructional practices, the *how* of teaching. Although I take a singular focus, teachers engaging in transformative work should be aware that the *what* (curriculum content) and the *how* (instructional practice) must work together to be effective in bringing about a change in a student's perceptions of the world. It matters what we teach, how we teach, and how we connect the two. If we use transformative methods to teach curriculum content that is not culturally relevant and does not reflect or encourage a diversity of perspectives, then change will not be initiated.

I use the term *transformative teaching* to represent a synthesis of different pedagogies. This is predicated on my belief that context affects the receptivity of the student to the pedagogical process. Context determines the space we have to make change and the kind of change we can make within a framework. Barndt (1989), who provides some very practical advice for those engaging in social activism, calls this "naming the moment." She argues that "there is a particular interrelationship or conjuncture of forces (economic, political and ideological) ... [that] shift from one moment to another ... When we plan actions, our strategy and tactics must take [these forces] into account" (Barndt 1989, 8). Given the continually changing power relationships among these forces, different moments provide us with different opportunities for change. In terms of classroom pedagogy, naming the moment alerts us to look at the dynamics of the class and at how these dynamics are framed by and interact with the educational institution and the social world.

Ropers-Huilman (1998) invokes Barndt's (1989) viewpoint by noting that there are a multiplicity of contradictory, overlapping, and fluctuating factors at play in the context of the classroom. hooks (1994) uses the term "engaged pedagogy" to describe the intermingling of different pedagogies. She insists that "engaged pedagogy recognize each classroom as different, that strategies must constantly be changed, invented, [and] reconceptualized to address each new teaching experience" (10–11). Framing pedagogy as a contextual practice means that no one pedagogy or instructional practice is right for all classrooms. As I teacher, I need to use varying blends of different pedagogies to support transformative teaching.

UNDERSTANDING AND USING TRADITIONAL PEDAGOGY

Pedagogy in the classroom is framed by the epistemological assumptions that underlie the concept of "knowing." The traditional pedagogy that dominates classroom teaching today is based on a positivist epistemology that conceptualizes "knowledge as a *thing* – essentially, as verifiable information born of scientific investigation" (Hinchey 1998, 39). A traditional pedagogy reproduces dominant class ideologies and power relationships by privileging certain knowledge and behaviours through course content, instructional practices, and a hidden curriculum. This form of pedagogy is "content-driven and teacher-centred ... [It privileges] the teacher's knowledge and experience as objective and superior to the students'" (Ironside 2001, 75). Traditional pedagogy in higher education lends itself to the lecture format, through which the teacher is able to impart expert and hegemonic knowledge to the student. The student gives feedback to the teacher through a pre-determined evaluative process. The primary focus in traditional pedagogy is on *what* the student learns, regardless of *how* it is taught.

To better understand how traditional pedagogy functions in a higher education classroom, I turn to Ehrensal's (2001) study of business education in the undergraduate classroom. He uses the concepts of "pedagogic authority," "pedagogic action," and "pedagogic work" developed by Bourdieu and Passeron (1990) to reveal how traditional pedagogy reproduces capitalist economic agendas through a hidden curriculum. Ehrensal (2001) argues that business education is designed to produce a form of "managerial habitus"; the acceptance and internalization of appropriate business practices and attitudes. Pedagogic authority determines who has the right and the responsibility to communicate the ideological message. In the traditional classroom there are two primary sources of pedagogic authority, the teacher and the textbook. Pedagogic authority is conferred on the teacher by the institution and by his or her own expert knowledge. The textbook brings to the classroom selected legitimated 'experts' in the subject area. Pedagogic authority in the traditional classroom frequently translates into pedagogic action in the form of a lecture. It is through the lecture and other forms of pedagogic action that the teacher's work is accomplished. Ehrensal (2001) describes the goal of this pedagogic work as the inculcation of "various forms of habitus that are both adaptive to and desired by the organizations with which [students] seek to find employment" (106). Case studies, experiential exercises, films, outside speakers, field trips, and internships are designed and used to model and reinforce acceptable behaviour and thinking in the business environment (Ehrensal, 2001). Those who internalize and exhibit "the appropriate habitus" (Ehrensal 2001, 108) are rewarded in school with good grades and in business with employment, advancement, and higher pay. Ehrensal (2001) shows how pedagogical practices that are frequently associated with transformative teaching are often used within a traditional pedagogy to reproduce the dominant ideological message. Teaching, as a form of activism, requires that the teacher be aware that an instructional practice may produce an illusion, rather than a reality, of fairness and inclusiveness. When adopting transformative pedagogical practices, the teacher must interrogate these practices and marry them with the course content in such a way that students gain insights that encourage them to question and seek change, rather than insights that reproduce dominant power relationships that maintain inequalities. When *what* is being taught does not fit with *how* it is taught, the possibility of change is negated.

UNDERSTANDING AND USING TRANSFORMATIVE PEDAGOGIES

Moving from a traditional pedagogy to pedagogies that support transformative teaching necessitates a change in the current practices of pedagogic authority, action, and work. I suggest that pedagogic authority should be shared between the teacher and the student, that pedagogic action include active rather than passive participation of the students in the learning process, and that pedagogic work include the building of tools that students can use to critically analyze and interrogate all types of knowledge. In this context, teachers and students work collaboratively to bring about change. Pedagogies that support transformative teaching provide an insight into how these changes might be theorized and practised in the higher education classroom.

A traditional pedagogy defines pedagogic authority as "formal power granted to individuals through institutional structures and relations" (Ng 1995, 132). Within this framework the pedagogic authority of the teacher cannot be shared with the students. However, if the power associated with pedagogic authority is framed as "a dynamic relationship negotiated in an interactional setting" (Ng 1995, 132), teachers have the opportunity to engage in pedagogic actions that make it possible to share power with students. The move towards a more egalitarian classroom changes the traditional concept of power, from a teacher's "power over," to a concept where all participants in the classroom have "power to" actively participate in pedagogical decisions and in the construction of knowledge (Hedin and Donovan 1989). Muzzin and Labreche (1999) note that fundamental to this notion of sharing power with students is the need to establish an environment of mutual trust and respect.

Teachers can share power in the classroom by involving students in decisions about classroom structure, procedures, and course requirements. Heinrich and Witt (1993) describe a classroom where "course objectives and methods of evaluation are mutually determined, not facilitator determined, and teaching approaches are interactive and experiential (120)." Hedin and Donovan (1989) emphasize the importance of using the first class in a course to foster two-way communication and to develop students as active participants in the learning process. They suggest that teachers make a clear statement about their own background and ideological standpoint and outline their expectations and proposed format for the class. marino (1997) advises teachers to "check with students about their expectations, their points of view [and] the criteria that they bring to a class, or an assignment, or a meeting" (50). Hedin and Donovan (1989) do not distribute their syllabus until the second class so that they can incorporate student input on what will be studied, how it will be studied, and how it will be evaluated into the course design. These teachers propose a combination of instructor-, peer-, and self-evaluation. Although it can be productive to use different forms of evaluation in transformative teaching, teachers should be cognizant of the problems associated with peer- and self-evaluation. Peer evaluation can be marred by a number of factors. These include personality differences and dysfunctional groups. Self-evaluation can be influenced by the personal expectations students have of themselves and the ways in which they perceive they have met those expectations. The teacher must monitor all forms of evaluation carefully, to take into account the different factors that may influence evaluation.

Transformative teaching starts with the assumptions that knowledge is socially constructed and that knowledge construction in the classroom should begin with the real-life situations and experiences of the students (Manicom 1992). Brookes (1992) believes that students should start theorizing from a familiar place. Students' understanding of their experiences is expressed in the "voice" they bring to the classroom. The teacher shares power by making space for and validating student voices. Differences and individuality can be respected by allowing students to express themselves and to discuss their experiences in non-traditional ways. Instructional practices can encourage innovation by including oral, written, and visual formats. Overall (1998) allows students to choose their own method of presentation, only requiring that it engage with the assigned reading and stimulate thoughtful discussion. In one of Overall's (1998) classes, student presentations included dance, music, art, questionnaires, excerpts from films, participatory theatre, group writing, small group discussion, role-playing, and poster-making.

It is not enough to bring student voice into the classroom. Students and teachers should also interrogate the values and beliefs that inform their understanding of their world (hooks 1994). Bell et al. (1999) use student journals to facilitate an "active integration of personal and theoretical material" (36). Students are asked to discuss "what they accept and reject, as well as their reasons why" (41). marino (1997) states that modes of expression, other than the written form, can also be valuable tools for exploring the self. She suggests that "photography, video and movies, music, and popular theatre have all been used effectively as tools for conscientization" (67–68).

"Learning is constructed not only through individual minds, but through the interactions of a community of knowers" (Maher and Tetreault 1996, 150). Student understanding of the self is built through relationships with others. Collaboration brings the individual student voice into contact with the voices of others through dialogue. Tarule (1996) describes this as "making knowledge in conversation" (280). Conversation can be facilitated through work in pairs and in small and large groups. Such pedagogic actions bring a plurality of voices into the classroom. This exposes students to the different ways of knowing that are a product of different social and cultural experiences. Awareness of different perspectives not only helps students to interrogate their own voice, but also to see how their voices are constructed through the eyes of another person (the "other"). "Before we can begin to claim to 'know' anything, we have to consider what a variety of others can tell us. No one else can give us a single accurate picture of what the world 'is,' what is 'important' in it: we have to construct our own understanding of the world for ourselves, basing it on a variety of sources" (Hinchey 1998, 59). Teachers may consider expanding the sources upon which knowledge construction is based by bringing guest speakers into the classroom and by taking students to venues outside the classroom. These instructional practices can be used to create a relationship between the everyday world and classroom knowledge, thus providing students with the opportunity to use classroom-generated knowledge to critique everyday practices in the social, political, and economic world.

Overall (1998) and hooks (1994) argue that for theory to be useful it must generate *praxis*. Praxis is an action-reflection-action cycle that facilitates transformation. The teacher can move towards an activist position by using praxis as a framework for pedagogical practices. The pedagogic work of the transformative teacher promotes activism when it facilitates change by equipping "others with the tools they need to deepen their understanding of their lives and to seek change *if* that is what they desire" (Hinchey 1998, 152). Hinchey (1998) identifies skills in cultural literacy that can be developed through transformative teaching. These include active and transformative reading; good expression and communication through speaking and writing; media, mathematical, and cultural literacy; and research and learning skills. Many of these skills are integral to most disciplines.

Fundamental to reflection and action in the classroom is the students' ability to critically question and interrogate their own position, the positions taken by others, and the structural and social elements of the dominant position that maintains inequalities. The transformative teacher must teach the students questions that will help them to move towards an activist position. Questions, which incorporate the notions of *how*, *why*, *so what*, and *what else* are useful for framing the transformative process. Gallagher (1997) suggests that the processes of knowledge creation can incorporate different perspectives by replacing the question "what happens next" with "what happens when" (28). marino (1997) believes that asking questions that start with "how would you know" (50) helps initiate a reflective and interrogative process. Pedagogic work profits from a simple questioning of everyday practices (Hinchey, 1998). Subsequent extension of this interrogation to the belief systems that inform everyday practices is a way of bringing an understanding from the local (the individual) to the global (how the individual is shaped by and shapes the social reality).

CONCLUSION

The preceding discussion has suggested that teaching in higher education can become a form of activism when transformative pedagogies, rather than a traditional pedagogy, are used in the classroom. Teachers become activists when they interrogate their own beliefs and pedagogical practices to determine the ways in which they maintain or challenge dominant forms of knowledge and existing power relationships. They become activists when they interrogate the theoretical frameworks that inform *what* they teach and *how* they teach, and when they interrogate the ways in which their pedagogical practices support or challenge the status quo.

Transformative teachers teach students to become activists when they use pedagogical practices that instill consciousness in students and expose the power relationships that maintain existing inequalities. Encouraging students to take a transformative stance and ask their own questions develops independent learning processes that enable them to take ownership of their learning. Students become activists when they use classroom knowledge to challenge dominant ideologies and practices in the social, cultural, and economic world.

In this discussion on transformative pedagogical practices, I have focused primarily on the notions of sharing power and of critical analysis. Although critical analysis is an important part of the transformative pedagogical process, it is still representative of the rationality that governs knowledge construction in higher education (Tisdell 2001). Both Ng (1998) and Tisdell (2001) posit that transformation is more readily accomplished if we move beyond the mind and rationality by incorporating other dimensions into classroom teaching. Ng (1998) states that current methods of teaching focus on the mind-intellect. She suggests that the "interdependence of the mind-intellect and body-spirit" (3) should be recognized in pedagogical practices. Tisdell (2001) recommends adding an affective and experiential dimension to critical analysis and to other pedagogical practices.

Transforming our understanding and creating new knowledge that can facilitate the challenging of power relations between dominant and oppressed groups require more than just critical analysis; they require transformation of the heart. Higher education has some responsibility to do this (Tisdell 2001, 160).

NOTES

1   Tisdell (2001, 148) defines positionality as "referring to how aspects of one's identity such as race, gender, class, sexual orientation, or ableness significantly affect how one is 'positioned' relative to the dominant culture."

REFERENCES

Barndt, D. 1989. *Naming the moment: Political analysis for action*. Toronto: Jesuit Centre for Social Faith and Justice.

Bell, S., Morrow, M., and Tastsoglou, E. 1999. Teaching in environments of resistance: Toward a critical, feminist, and antiracist pedagogy. In *Meeting the challenge: Innovative feminist pedagogies in action*, M. Mayberry and E. Cronan Rose, eds., 23–46. New York: Routledge.

Brookes, A.L. 1992. *Feminist pedagogy: An autobiographical approach*. Halifax, Nova Scotia: Fernwood Publishing.

Ehrensal, K.N. 2001. Training capitalism's foot soldiers: The hidden curriculum of undergraduate business education. In *The hidden curriculum in higher education*, E. Margolis, ed., 97–113. New York: Routledge.

Ellsworth, E. 1992. Why doesn't this feel empowering? Working through the repressive myths of critical pedagogy. In *Feminisms and critical pedagogy*, C. Luke and J. Gore, eds., 90–119. New York: Routledge.

Gallagher, K. 1997. Essentially different: Creative drama and the politics of experience in girls' education. *National Drama*, 21(2), 17–31.

Hedin, B.A. and Donovan, J. 1989. A feminist perspective on nursing education. *Nurse Educator*, 14(4), 8–13.

Heinrich, K.T. and Witt, B. 1993. The passionate connection: Feminism invigorates the teaching of nursing. *Nursing Outlook*, May/June, 117–24.

Hinchey, P.H. 1998. *Finding freedom in the classroom: A practical introduction to critical theory*. New York: Peter Lang Publishing.

hooks, b. 1994. *Teaching to transgress: Education as the practice of freedom*. New York: Routledge.

Ironside, P.M. 2001. Creating a research base for nursing education: An interpretative view of conventional, critical, feminist, postmodern, and phenomenologic pedagogies. *Advances in Nursing Science*, 23(3), 72–87.

Kenway, J. and Modra, H. (1992). Feminist pedagogy and emancipatory possibilities. In *Feminisms and critical pedagogy*, C. Luke and J. Gore, eds., 138–65. New York: Routledge.

Maher, F.A., and Tetreault, M.K. 1996. Women's ways of knowing in Women's Studies, feminist pedagogies, and feminist theories. In *Knowledge, difference, and power: Essays inspired by women's ways of knowing*, N. Goldberger, J. Tarule, B. Clinchy and M. Belenky, eds., 148–70. New York: Basic Books.

Manicom, A. 1992. Feminist pedagogy: Transformations, standpoints, and politics. *Canadian Journal of Education* 17(3), 365–389.

marino, d. 1997. *Wild garden: Art, education, and the culture of resistance*. Toronto, ON: Between the Lines.

Muzzin, L. and Labrèche, D. 1999. *Mutually reflecting on teaching and personal transformation*. Paper presented at the 16th Qualitative Analysis Conference: The Interdisciplinary Study of Social Processes, May 13–16, University of New Brunswick, Fredericton, New Brunswick.

Ng, R. 1995. Teaching against the grain: Contradictions and possibilities. In *Anti-Racism, feminism, and critical approaches to education*, R. Ng, P. Staton and J. Scane, eds., 129–52. Westport, CT: Bergin and Harvey.

Ng, R. 1998. *Is embodied teaching and learning critical pedagogy? Some remarks on teaching health and the body from an eastern perspective*. Paper presented at the AERA annual meeting, April 13–17, San Diego, CA.

Overall, C. 1998. *A feminist I: Reflections from academia*. Peterborough, ON: Broadview Press.

Ropers-Huilman, B. 1998. *Feminist teaching in theory and practice: Situating power and knowledge in poststructural classrooms*. New York: Teachers College Press.

Tarule, J. M. 1996. Voices in dialogue: Collaborative ways of knowing. In *Knowledge, difference, and power: Essays inspired by women's ways of knowing*, N. Goldberger, J. Tarule, B. Clinchy, and M. Belenky, eds., 274–99. New York: Basic Books.

Tisdell, E. J. 2001. The politics of positionality: Teaching for social change in higher education. In *Adult education and the struggle for knowledge and power in society*, R.M. Cervero, A.L. Wilson, and Associates, 145–63. San Francisco, CA: Jossey-Bass.

ACKNOWLEDGMENTS

Special thanks to Margrit Eichler, Peggy Tripp, and Linda Muzzin for their constructive comments as I was writing this paper.

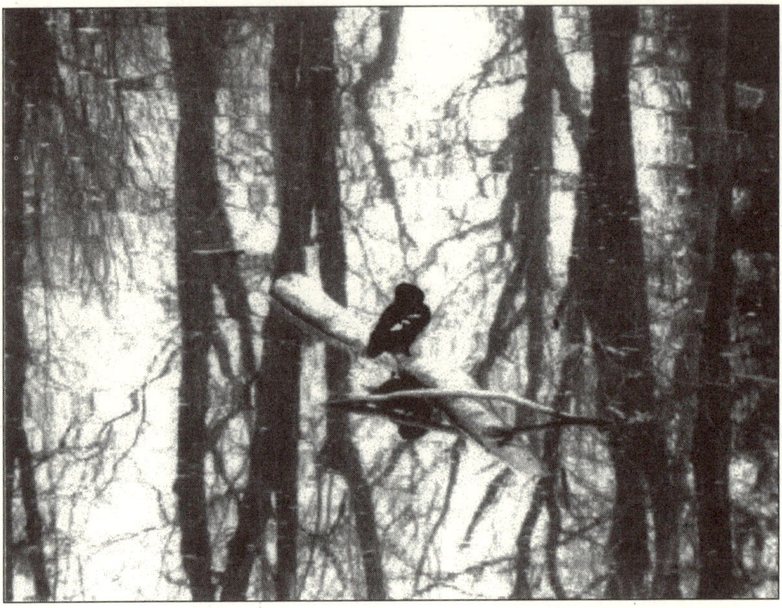

7

Science, Environment, and Women's Lives: Integrating Teaching and Research

MARIANNE GOSZTONYI AINLEY

During the past 25 years, my research has focused on the histories of Canadian science and environment. My ongoing investigations have, in turn, motivated my interest in developing undergraduate teaching material on "women, science, and technology" and "women, power, and environments" both from historical and contemporary perspectives. Later, these investigations led to my constructing graduate seminars around the themes of "feminist perspectives on science and technology," and "gender, power, and environmental problems." Both my teaching and research have contributed to feminist studies of science and environments in a variety of ways.

In 1977, I entered graduate school to study the history and sociology of science, after having worked in scientific/technical environments (industry, government, and academe) for two decades. My own experiences as an invisible woman chemist, combined with my activities as an amateur ornithologist, led to a scholarly interest in those on the margins of the Western scientific community. My work was considered *avant-garde* by some of my professors and marginal to the main topics of contemporary inquiry by others. At the time, most history and sociology of science research centred on Western science, eminent male scientists, and scientific institutions. It focused on the laboratory-based sciences such as nuclear physics, molecular biology, chemistry, and some sub-fields of biology and psychology. This type of research considered the extent of professionalization as a sign of a science's maturity. By contrast, less visible scientists and institutions, and the field-oriented sciences (such as some areas of botany, zoology, and ecology) were neglected. Only a handful of scholars wrote about the "contribution of the amateur" in astronomy, archaeology, and field ecology (Lowe 1976; Stebbins 1980).

As someone who had participated in ornithological investigation as a non-professional, I knew that many others (also called amateurs, volunteer investigators, or avocational scientists), had been involved in long-term field studies. Yet, the above trends excluded them from scholarly investigation in science, thereby trivializing their contributions. This led to my own research on the professionalization of North American ornithology. In my master's thesis, I showed that in the second half of the twentieth century, some sciences remained incompletely professionalized, and that marginal scientists had a more central role to play in science than the contemporary literature on the subject led us to believe. While I did no gender analysis in this work, I dealt with power relations in an incompletely professionalised science. This was my first foray into writing about a seemingly marginal segment of the scientific community (Ainley 1980).

From 1980 to 1985, I was at McGill University working on my doctorate. My dissertation on the transformation of "Natural history to avian biology: Canadian ornithology, 1860 to 1950" focused on natural history explorations, government science, changing trends in science, tensions between the field and laboratory oriented sciences, the contribution of non-professionals to an increasingly professionalized and institutionalized scientific field, and issues of environments – both natural and institutional. My work also dealt with research funding, government priorities, and scientific careers.

Women were among the ornithologists I studied, but I still lacked a gender analysis. This may be surprising because it was during this period that scholars began to focus on women and science. In the United States, women scientists formed women's caucuses within their own disciplinary associations. In Canada women scientists, such as Maggie Benston, Hilda Ching, and Rose Sheinin were among those who studied women's status in science and organized the First National Conference for Women in Science, Engineering and Technology in Vancouver in May 1983 (Benston 1982, 1983; Ching 1983; Scott and Ferguson 1982; Sheinin 1984).

While my graduate training did not prepare me to deal with gender issues, the publications of American historian of science Margaret Rossiter (1980, 1982) opened my eyes. I became interested in the history of women scientists and, in 1983, in Hungary, attended the First International Interdisciplinary Conference on Women in the History of Science,

Technology, and Medicine in the Nineteenth and Twentieth Centuries. I was the only Canadian delegate at this conference; my talk was about American women ornithologists and none of the other historians had studied the Canadian situation. This experience was a major turning point for me. From my own background I already knew that women scientists worked in government, academe, and industry. I wondered why there was nothing in the literature about them. In 1984, with a grant from the Canadian Research Institute on the Advancement of Women (CRIAW), I embarked on the first stage of my research program on a history of Canadian women in science. During the following decade, a SSHRC postdoctoral fellowship and several grants allowed me to enlarge the scope of my research. Concurrently, I worked on a feminist scientific biography of University of Alberta zoologist, ecologist, and conservationist Dr William Rowan (1891–1957) (Ainley 1993).

These seemingly disparate research areas were, in reality, interwoven projects. William Rowan's extensive correspondence with other scientists (both women and men) from 1916 to 1957 illustrated important issues in these scientists' lives. Among them were the powerlessness of academics vis-à-vis university presidents, and work-related stresses due to large classes, long working hours, inadequate or non-existent scientific equipment, and lack of funding. I gained insight into the role of gender and other power relations in scientific environments from reading these letters. Working simultaneously on the biography of a famous Canadian man of science and on a history of lesser-known women scientists in the first half of the twentieth century enabled me to differentiate between the general problems faced by most scientists and those additional ones which were gender-specific. This was particularly important because we knew little about the historical situation of women and about gender relations in Canadian science.

The history of Canadian science and the history of women in science were relatively new fields of scholarly inquiry and the first books on Canadian science published during the second wave of feminism did not mention women. When, in the 1980s, a few historians began to investigate the gendered history of Canadian science, they found that science books, reference books, and general histories of the various scientific fields ignored or minimized women's role in science and that women scientists were, for the most part, excluded from the written record. In fact, beyond sporadic names and facts, we knew nothing about the women, and not much about the men, whose knowledge and discoveries contributed to the geopolitical entity we now call Canada. Clearly, there was room for my proposed historical research on Canadian women and scientific work.

After a frustrating year as a post-doctoral fellow at McGill University (1985–86), I obtained a Women and Work SSHRC Strategic grant and a research affiliation with the Simone de Beauvoir Institute at Concordia University. I began to travel across the country to work in various archives and interview women scientists. My colleagues at the Simone de Beauvoir Institute provided emotional and intellectual support and periodically challenged my assumptions. This was crucial because neither my own work experience nor my graduate training provided me with the tools required for feminist research. While I still had an incomplete understanding of the scope and complexity of the research I had undertaken, I knew that my own historical research could create a new dimension in the history of science. Providing historical data would help demystify science and the history of science, and the new historical documentation could lead to science policy changes. My investigations of the histories of several scientific fields and gender relations in Canadian science ultimately resulted in a new feminist story-telling that deals with gender and other power relations within the Canadian scientific community. This history of Canadian science includes both women and men, and demonstrates that the various phases of the history of Canadian science were not as clear-cut as previously believed and that not all the scientists were from the dominant Western culture, nor the dominant sex.

Historical research is labour intensive, particularly on marginalized groups. During the next few years, I spent hundreds of hours searching for information in numerous archives and interviewing scientists. I learned both from the scientists' own insights on their experiences and from their perspectives on disciplinary and institutional contexts. Despite all the information I found there was still not enough Canadian material for the women and science course that I developed for Concordia University in 1988. And so, while continuing my research on Canadian women natural scientists, I also decided to edit a collection of essays and sent out invitations to a variety of feminist scholars asking them to contribute to this book.

Since the publication of *Despite the odds: Essays on Canadian women and science* (Ainley 1990), my research has expanded in both time period and scope. Based on my historical data, it is now possible to see historical trends as well as individual, disciplinary, and institutional differences. While my doctoral research had provided me with ample documentation on mostly male scientists, and an androcentric and Eurocentric context, the women and science research program produced new information and new insights. I benefited from the work of other historians (on women, Canada, and First Nations) and from anthropologists and ethno-scientists conducting oral histories in Aboriginal communities. Based on interdisciplinary research, my own and that of others, my interpretation of the history of Canadian science differs from male stream research on the subject. In contrast to previous historical writings, I argue the following eight points 1) Beginning in the seventeenth century, Indigenous people transferred their knowledge of the environment to many men and a few women trained in the European scientific tradition. 2) Early-nineteenth-century British women such as Lady Dalhousie (1786–1839), Anne-Marie Perceval (1790–1876), and Harriet Sheppard (d.1877) were active agents in an emerging Canadian scientific community (Ainley 1990, 1997). 3) Women such as Catharine Parr Traill (1802–1899), Annie Jack (1839–1912), and Eliza Jones (d.1903) played important roles in science education and research throughout the nineteenth century. They did fieldwork, published popular works on science, and influenced three generations of immigrant and Canadian-born readers (Ainley 1997). 4) The late-nineteenth-century convergence of the institutionalization of science and women's entry into higher education initially opened up science as a 'career' – now a problematic term – for white middle-class Anglo-Canadian women. Between 1890 and 1970, hundreds of women scientists in Canada found employment at private and public schools, universities, industrial and hospital laboratories, and the scientific branches of the Canadian government, such as the Geological Survey, the Department of Agriculture, and the National Research Council of Canada. Many other women (with or without university degrees) contributed to science as voluntary scientists or so-called amateurs (Ainley 1990). 5) Most Canadian scientists before 1945 came from the dominant British-Canadian culture. Second-generation Canadian women from non-British backgrounds entered science after 1945, as did European-trained women scientists. By contrast, although French-Canadian nuns have contributed to science since the seventeenth century, few secular French-Canadian women studied science until the 1960s. In the meantime, science became an increasingly prestigious occupation for French Canadian men. 6) Competition for scarce scientific posts increasingly resulted in the creation of a niche of "women's work" – that is, women either remained in the lower echelons of a stratified Canadian scientific community or moved to sub-fields considered suitable for women. Moreover, socio-economic and ethnic backgrounds were more likely to adversely affect the professional opportunities of women than men. 7) After 1960, legalized and improved contraception could delay child bearing, but the career versus marriage/motherhood conflict continued. Women who interrupted their career for childrearing never caught up with their male colleagues. By contrast, men who wanted to have a family rarely suffered from the lack of a social support system (legislations or policies concerning maternity leave). 8) Although more women have obtained positions in science since 1960, most of them have had fewer grants and less pay than men of the same age and same graduate training. Lack of peer recognition and exclusion from the informal network of science has continued to work against women having professional opportunities similar to men's.[1]

What about my teaching? In retrospect I see that my teaching and my research have become broader and more interdisciplinary during the past dozen years. I use feminist perspectives in my interdisciplinary courses and draw on materials from history, history of science, anthropology, sociology, geography, environmental studies, literature, women's studies, and science. I also introduce students to a broad view of science that includes the natural, social, and applied sciences as well as Indigenous knowledge systems. My favourite definition of science is by the American zoologist Marion Namenworth (1986, 19), who wrote, "Science is a system of procedures for gathering, verifying, and systematizing information about reality," because this definition is broad rather than narrow and restrictive. It acknowledges the fact that different people have different realities, and that there are many ways to gather, verify, and systematize information. Thus it opens the way to considering women's knowledge in general and Indigenous women's knowledge in particular along a broad spectrum of scientific knowledge and activity. I use the same perspective in my ongoing research, which I integrate into my teaching, particularly in four upper-level courses, two of them in the Gender Studies MA program. My ongoing reflections on the interdisciplinary course material on women, science, and environments have become, in turn, relevant for my research.

In the undergraduate and graduate Women and Science courses at my university, we explore historical trends and contemporary concerns regarding women, science, and technology; deconstruct prevailing stereotypes of science and scientists; compare traditional First Nations knowledge and Western science; analyze the interrelationships of gender, race, and science; and study science as a social activity. We also discuss the changing face of science and the possibilities for a community-based, integrated science. In addition, the graduate course deals in more detail with the origins of Western science; the multicultural aspects of science and the history of science; the cultural construction of science and technology; male stream and colonial science; women/nature/science; feminist and postcolonial perspectives on gender/race/nature; scientific literacy and its gender implications; the ethnography of different scientific fields; and finally, new directions in the study of women, gender, and science.

The two environmental courses in my program that deal with women/gender, power, and environments focus on women's relationships with their environments from prehistoric times to the present; discuss gendered environmental histories (including family history), issues, and concerns across cultures; as well as Western expansion, warfare, and internal environments (including "chilly climates") from feminist perspectives. Students develop environmental awareness through the course material and personal experience, and create strategies to challenge existing power relations that are detrimental to our present and our future. They learn to use a critical feminist lens to analyze power relations that cause environmental degradation in modern societies.

The issues that we discuss in class are all relevant to research on women, science, and environments in the former British colonies and other parts of the world. This has contributed to my extending my previous research program during the past two years to carry work on women/gender science and environments a step further. As a feminist scholar I feel that I need to re-visit sources I first looked at 10 to 20 years ago and re-think their implications and my own assumptions. I and other historians of science need to do more comparative research, particularly on the transfer of knowledge from Indigenous people to Western science.

My new research program, "Re-explorations: New perspectives on gender, environments, and the transfer of knowledge in nineteenth and twentieth century Canada and Australia" is the first by a feminist historian of science studying either the process of knowledge transfer from Indigenous women to women of European extraction or the dissemination of traditional environmental knowledge into Western science. Because in most of the Western world neither women's knowledge nor Indigenous environmental knowledge has been sufficiently considered, a re-exploration of gender, science, and environments in the former British colonies from feminist post-colonial perspectives is a much needed and challenging field of inquiry.

In my previous publications, I argued, as have many other authors in recent works on women naturalists, that the British tradition of natural history opened the way for marginal people (women and amateurs) to contribute to the field-oriented sciences well into the twentieth century (Gates and Shteir 1997; Pratt 1992; Kuklick and Kohler 1996). From the writing of anthropologists and ethno-historians we know that Indigenous peoples' traditional environmental knowledge has been practised in the field and acquired both through apprenticeship to Elders and by first-hand field experience. Some of this knowledge found its way into Western science. I intend to study how, when, and where this knowledge was transferred and later disseminated – or in some cases ignored and even suppressed by the male scientific establishment.

Clearly, my current research forms both an integral part and an extension of my long-term research program on Canadian women and science. It focuses on nineteenth and twentieth century colonial encounters in Canada and Australia between Western and Indigenous women, and knowledge transfer from Indigenous women to Western women travellers, settler-naturalists, and later scientists. While Indigenous-White relations differed in Canada and Australia, White women in both countries benefited from imperial expansion, Anglo settlements, and the British tradition of travel and natural history writing. In North America and Australia, they collected natural history specimens, and wrote books and articles primarily for a general reading public and, to a lesser extent, for botanists and zoologists. They encountered varied environmental and social conditions in the different settler colonies and this influenced their perspectives, studies, and writings. We do not yet know to what extent Indigenous women influenced their 'discoveries.'

From the works of Australian scholars, we know that British and other European women visitors to Australia, such as Elizabeth Gould (1804–1841) and Lady Jane Franklin (1791–1875) contributed to Western science with their collections, illustrations, and other activities. Privileged women settlers, such as Georgiana Molloy (1805–43) and Louisa Atkinson (1834–72) observed, collected and described the natural history of parts of Australia (Moyal 1986, Maroske 1993). We do not yet know to what extent their association with Aboriginal women influenced their studies. We do know that male surveyors in Australia relied on Aboriginal people for food, water, and shelter, and included Aboriginal men and women in their exploring parties.[2] Australian anthropologists Sandy Toussaint (1999) and Diane Bell (1993) and ethno scientists, such as Jennifer Isaacs (1987) and Peter Latz (1999), have done much during the past few decades to retrieve Aboriginal women's knowledge. Their findings will be integrated into my interdisciplinary research on gender, environments, and the transfer of knowledge.

Such research will help broaden our view of what constitutes science, and how and where science is learned and practised. It will contribute to cutting-edge scholarship on the history of colonizer and colonized women in the nineteenth and twentieth centuries, to histories of gender and science, environmental history, indigenous histories, post-colonial science studies, and the histories of science in settler societies.

From the foregoing, it is clear that both my own research and others' scholarship on women, science, and environments have undergone major changes in the last quarter century. The early practitioners of the field were women scientists trained in a Western scientific milieu, carrying the conceptual baggage of Western science, but beginning to realize the negative impact of this tradition on women, science, and environments. Since then, the field has undergone many changes and has become more grounded in the realities of women's lives. Many feminist historians around the world have provided empirical data on women scientists. These have contributed to a new storytelling in the history of science and have led to some science and educational policy changes. When I began my research on the Canadian situation, I too carried around the conceptual baggage of a Western-trained scientist and historian of science. After I developed a feminist perspective and recognition of the contributions of Indigenous knowledge to Western science, I expanded the scope of my research, questioned malestream histories of Canadian science that excluded women and Indigenous people, and strove to contribute to a new and more integrated history of Canadian science. My teaching and graduate supervision enabled my students at two universities to develop a critical perspective on the history and on current practices of Western science and issues regarding Indigenous and colonial science. It improved their scientific literacy and alerted them to the importance of engaging in activism regarding gender, science, and environments. All of them will be able to use their new knowledge in their everyday lives. Many will also contribute to new interdisciplinary areas of scholarship about gender, science, and environments, and will teach and do research as if the world mattered.

NOTES

1 Fourteen panel discussions with women scientists, 1988–2002, at Concordia, Carleton, and the University of Northern British Columbia.

2 The British surveyor, Major Thomas Mitchell was among them. See Ryan (1996).

REFERENCES

Ainley, M.G. 1980. *La Professionalisation de l'ornithologie Américaine*, unpublished MSc thesis, Université de Montréal.
– ed. 1990. *Despite the odds: Essays on Canadian women and science*. Montreal: Véhicule Press, 1990.
– 1990. Last in the field? Canadian women natural scientists. In *Despite the odds*, ed. M.G. Ainley, 25–62. Montreal: Véhicule Press.
– 1993. *Restless energy: A biography of William Rowan, 1891–1957*. Montreal: Véhicule Press.
– 1997. Science in Canada's backwoods: Catherine Parr Traill, In *Natural eloquence: Women reinscribe science*, eds. B.T. Gates and A. B. Shteir, 79–97. Madison: University of Wisconsin Press.
Bell, D. 1993. *Daughters of the dreaming*, 2nd ed. Minneapolis: University of Minnesota Press.
Benston, M. 1982. Feminism and the critique of scientific method. In *Feminism in Canada*, eds., A. Miles and G. Finn, 57–76. Montreal: Black Rose.
– 1983. Technology in the workplace: Chipping away at women's work, *Herizon* 1(7), 19–24.
Ching, H.L. ed. 1983. *Proceedings of the first national conference for Canadian women in science and technology*. Vancouver, May 20–22.
Gates, B.T. and A.B. Shteir, eds. 1997. *Natural eloquence: Women reinscribe science*. Madison: University of Wisconsin Press.
Isaacs, J. 1987. *Bush food: Aboriginal food and herbal medicine*. Sydney: Lansdowne.
Kuklick, H. and R.E. Kohler, eds. 1996. Science in the field. *Osiris* 11 (theme issue).
Latz, P. 1999. *Pocket bushtucker*. Alice Springs: IAD Press.
Lowe, P.D. 1976. Amateurs and professionals: The institutional emergence of British plant ecology. *Journal of the Society for the Bibliography of Natural History* 7, 517–32.
Maroske, S. 1993. The whole great continent as a present: Nineteenth-century Australian women workers in science. In *On the edge of discovery: Australian women in science*, ed. F. Kelley, 13–34. Melbourne: The Text Publishing Company.
Moyal, A. 1986. The feminine touch. In *A bright and savage land: Scientists in colonial Australia*, ed. A. Moyal, 106–17. Sydney: Collins.
Namenworth, M. 1986. Science seen through a feminist prism, In *Feminist approaches to science*, ed. R. Bleier, 18–41. New York: Pergamon Press.

Pratt, M.L. 1992. *Imperial eyes: Travel writing and transculturation.* London and New York: Routledge.

Rossiter, M. 1980. Women's work in science, 1880–1910. *Isis* 71:381–398.

– 1982. *Women scientists in America: Struggles and strategies to 1940.* Baltimore: Johns Hopkins University Press.

Ryan, S. 1996. *The cartographic eye: How explorers saw Australia.* Cambridge: Cambridge University Press.

Scott, J.P. and J. Ferguson. 1982. *Who turns the wheel?* Ottawa: Science Council of Canada.

Sheinin, R. 1984. Women in science: Issues and actions. *Canadian Woman Studies/Cahier de la Femme* 4/5, 70–7.

Stebbins, R. A. 1980. Avocational science: The amateur route in archaeology and astronomy. *International Journal of Comparative Sociology* 21, 34–47.

Toussaint, S. 1999. *Phyllis Kaberry and me: Anthropology, history, and aboriginal Australia.* Melbourne: Melbourne University Press.

SECTION TWO

Problematization of Dominant Realities

Colonial Marinade

8

You Can't Be the Global Doctor If You're the Colonial Disease

MARIE BATTISTE

ENDANGERED INDIGENOUS KNOWLEDGE:
GLOBAL CONCERNS

The United Nations declared 2001 the Year of Dialogue among Civilizations, providing an exemplary opportunity for Canadian educational institutions to confront the ethics, methodologies, and lessons of Indigenous knowledge and heritage. This was a starting point for the long-delayed but necessary dialogue on a just construction of globalization, which is the latest in a series of artificially created concepts that are often referred to but seldom defined. Since globalization continues patterns of imperialism privileging Eurocentric thought, Indigenous communities find themselves still locked in the continuing struggle to overcome the destructive effects of colonization.

This paper focuses on the contributions Indigenous knowledge can make to educators and educational institutions in drawing upon and centring Indigenous knowledge. I will discuss how centring the dynamic foundations of Indigenous knowledge in their curricula create possibilities related to the growing awareness of our endangered environment, and how a learning environment contaminated with Eurocentric knowledge limits this awareness.

Post-colonial theory helps us unravel the colonial mentality that has endangered and subjugated peoples around the world. It raises our consciousness, develops our resistance, and helps us engage in transformative action. The decolonization of education is on the agendas of many forward-thinking individuals who view the earth as significant enough to warrant protection. These are people who believe that it is worthwhile to find and centre knowledge that embraces holistic principles of respect. Indigenous knowledge may be endangered, but is still available to those who wish to acquire and benefit from it. It must, however, be accessed in ways that are respectful to those who have nurtured it in the past and will sustain it for the future.

The neglect and destruction of Indigenous knowledge is a widely felt historical legacy. All around the surface of the earth, Indigenous peoples live in communities where they acquire, develop, and sustain relationships with each other and with their environments. By building relationships with the land and its inhabitants, they come to understand the forces around them. Each generation then passes their knowledge and experience of the social and cultural contexts of their ecological origins to succeeding generations. They transmit their knowledge through their languages and through many diverse ceremonies and traditions. These cultural forms are the fundamental sources of Indigenous knowledge.

Indigenous knowledge is not a peculiar subset of knowledge, but a diverse array of knowledges that are distinctive to different peoples and to their varied environments. It is everywhere, and it emanates in many diverse cultural forms throughout the world. Early modern science considered Indigenous knowledge to be peculiar to culture, and many anthropologists and linguists spent much time and energy discovering the common and exotic aspects of different Indigenous cultures. But social scientists have failed to capture the breadth of Indigenous knowledge and have confused its nature with its complex and interrelated parts. Over the years, it has been marginalized and often discarded or suppressed.

The appropriation of Indigenous knowledge is a more recent problem. In 1987, at a benchmark event for sustainable development in Sweden, the Brundtland Commission declared Indigenous knowledge to be part of a sustainable ecological knowledge that has existed for thousands of years. Since that conceptualization, traditional ecological knowledge, which is intimately related to the rest of Indigenous knowledge, has gained credibility in international research, communities, and law. The recognition that Indigenous knowledge is beneficial and valuable to modern society has created an impressive scientific reaction as the international community pursues its quest for sustainable economies and lifestyles. The search to understand Indigenous knowledge is part of a new paradigm dedicated to maintaining our earthly environment. The knowledge Indigenous peoples have gained of plant and animal behaviour, as well as their sustainable management of natural resources, has inspired new partnerships with and interest in Indigenous peoples themselves. However, despite the fact that Indigenous knowledge is increasingly being seen as a valuable tool in the areas of modern medicine and environmental concerns, for Eurocentric thinkers, it remains a mysterious and uncharted realm.

This new interest in Indigenous knowledge has raised issues about who owns intellectual and cultural property, and it has fuelled political confrontations between Indigenous and non-Indigenous peoples around the globe. One example of these confrontations is being played out in the rainforests of South America. Pharmaceutical companies avoid the costs of research and development by going directly to Indigenous peoples for their knowledge of medicinal plants. They then appropriate this knowledge, patent it, and sell it back to Indigenous peoples. The result is that the international community is faced with yet another form of global racism that threatens many Indigenous peoples. This racism assumes the cultural superiority of the pharmaceutical companies over colonized peoples, empowering the scientific community to commodify the products of Indigenous knowledge without Indigenous peoples' collective consent and cooperation, and without adequate compensation or consideration of the impact such commodification may have on the collective that developed this knowledge. The appropriation of Indigenous knowledge by Western interests is thus a growing area of concern for Indigenous peoples.

Western thought uses Indigenous knowledge for its own purposes, and the hierarchical structure of Western knowledge, which rates some kinds of knowledge as more valuable than other kinds of knowledge, has a destructive impact on many Indigenous peoples. Feminist epistemologist Elizabeth Minnich (1990) has called the structure of Western knowledge "hierarchical invidious monism." I use the term "Eurocentric," although others have pointed out the diversity of Europe as well as the diversity of those who have come under its dominion. So whether it is called Anglocentric, Western, Eurocentric, or hegemonic colonial knowledge (or Minnich's "hierarchical invidious monism"), the structure manifests itself in common forms and has common assumptions that support it.

To understand why Indigenous knowledge is marginalized in Western thought is to unravel Eurocentrism, at the heart of which is the theory of Eurocentric diffusionism (Blaut 1993). As the dominant artificial context for knowledge during the last five centuries, Eurocentrism postulates the superiority of Europeans over non-Europeans. Eurocentrism is not just an opinion or attitude that can be changed by some multicultural or cross-cultural exercise, for Eurocentrism is an integral foundation of all dominant scholarship, opinion, and law. As an imaginative and institutional context, Eurocentrism is the dominant consciousness and order of contemporary life. It is a consciousness in which all of us have been marinated.

Eurocentrism is an ultra-theory in modern thought. It is the context for many smaller historical, geographical, psychological, sociological, and philosophical theories, all of which can be seen as integral parts of Eurocentric 'diffusionism.' Diffusionism in its classic form divides the world into two categories. One category (Greater Europe, Inside) is historical, invents, and progresses; the other category (non-Europe, Outside) is ahistorical, stagnant, and unchanging, and receives progressive innovations by diffusion from Europe. From this base, diffusionism attributes the difference between the two peoples represented in this dialectic to some intellectual or spiritual factor, some characteristic of the 'European mind,' the 'European spirit,' or 'Western Man.' It is this characteristic that supposedly leads to creativity, imagination, invention, innovation, rationality, and a sense of honour or ethics – in other words, to 'European values.' The reason for non-Europe's non-progress is said to be a lack of this intellectual or spiritual factor. Diffusionism asserts that non-European people are empty, or partly so, of 'rationality' – that is, of ideas and proper spiritual values (Blaut 1993).

All Eurocentric scholarship is diffusionist since it axiomatically accepts the Inside-Outside model, the notion that the world as a whole has one permanent centre (Europe) from which culture-changing ideas tend to originate, and a vast Indigenous periphery that changes as a result (mainly) of diffusion from that single centre. Modern Eurocentric thought does not claim to be a privileged norm. This would be an argument of cultural relativism, which asserts that values come from specific cultural contexts. Instead, Eurocentric thought has always claimed to be universal. Eurocentric thinkers have often used this claim of universality to project Eurocentric beliefs onto other cultures that possess different world views or localized knowledge.

Universality is one aspect of diffusionism, and claiming universality often means aspiring to domination. Universality underpins cultural and cognitive imperialism, which establishes a dominant group's knowledge, experience, culture, and language as the universal norm. Dominators (or colonizers) reinforce their culture and values by bringing the oppressed and the colonized under their expectations and norms. Because Eurocentric colonizers consider themselves to be the ideal model for humanity and carriers of superior culture and intelligence, they believe they can judge other people and assess their competencies. In short, Eurocentric academics believe they have the power to research and interpret differences, and this belief has shaped the institutional and imaginative assumptions of colonization and modernism. Using the strategy of differences, they believe they have the privilege of defining human competencies and deviancies, as well as the authority to impose their tutelage – through education – over Indigenous peoples.

Given the assumed normality of the dominators' values and identity, the dominators construct the differences of the dominated as inferior and negative and those of the dominators as superior and positive. One of the foremost theorists of colonization, Albert Memmi (1965), explains the development of this binary consciousness and the society of the immigrant-colonizer and the Indigenous-colonized, which the colonized have to accept if they are to survive. This binary consciousness has been used to justify the separation of Indigenous peoples from their ancient rights to the land and its resources, and the transfer of wealth and productivity to the colonialists and the "mother" country.

Typically, to succeed in creating the belief that their world view is universal and therefore objective, colonizers must erase Indigenous memories and knowledge. Without significant exception, the universal discourses of Eurocentric thought force silence on Indigenous peoples. When the colonizers arrive, the Indigenous peoples lose their languages, their cultural continuity, their children, and their histories. The ensuing silence strips Indigenous peoples of their heritage and identities, while the Eurocentric education and legal systems induce a collective amnesia that alienates Indigenous peoples from their Elders, from their linguistic consciousness, and from their order of the world. According to diffusionism, only superior peoples can be the agents of progress, either by the will of God or by the laws of nature. European learning is established as the universal model of civilization, which must be imitated by all other groups and individuals. Thus, Eurocentrism monopolizes history, progress, and interpretation. It attempts to be the central agent of progress, the creator of knowledge and the doctor to disease as well as social and economic ills.

Eurocentrism creates a strategy of differences that leads to racism, which then allows the colonialists to assert their privileges while exploiting Indigenous people in an inhuman way. Memmi outlines this strategy:

"Racism is the generalized and final assigning of values to real or imaginary differences, to the accuser's benefit and at his victim's expense, in order to justify the former's own privileges or aggression" (1969, 185). Memmi proceeds to identify four strategies used to maintain colonial power over Indigenous people: 1) stressing real or imaginary differences between the racist and the victim; 2) assigning values to these differences, to the advantage of the racist and the detriment of the victim; 3) trying to make these values absolutes by generalizing from them and claiming that they are final; and 4) using these values to justify any present or possible aggression or privileges.

Linguistic ecologist Robert Phillipson (1992) demonstrates that many of the basic terms used in English and French analyses of Indigenous peoples and their knowledge are ideologically loaded with the tenets of Eurocentric diffusionism. The words used to describe Indigenous peoples reflect a European way of conceptualizing the issues and tend to reinforce colonial myths, racism, and stereotypes. Many English concepts establish a pattern of Eurocentric self-exaltation that creates an idealistic image of itself and devalues Indigenous knowledge. For example, European nations have 'languages' whereas tribes have 'dialects.' European nations have 'knowledge' whereas tribes have 'culture.' These perspectives shape the broad understanding of non-Eurocentric knowledge systems. The drafters of most international documents and academic definitions are unaware of the Eurocentric biases and traditions that are integral to and hidden in their language systems.

Eurocentric education policies and attempts at assimilation have contributed to major global losses in Indigenous knowledge. In Canada, the Royal Commission on Aboriginal Peoples (RCAP) wrote an extensive report that illustrated the massive damage done to all aspects of Aboriginal peoples' lives. The 1996 report was the result of a huge mobilization of Canadian scholars and public servants in an effort to unravel the effects of generations of exploitation, violence, marginalization, powerlessness, and enforced cultural imperialism on Aboriginal knowledge and peoples. The RCAP's conclusions and recommendations reflect the broad consensus of 150 distinguished Canadian and Aboriginal scholars, and the deliberations of 14 policy teams comprising senior officials and diverse specialists in government and politics (RCAP, 1996, 5:296–305).

The report creates a post-colonial agenda for transforming the traumatic relationship between Aboriginal peoples and Canadians. It affirms how the false assumption of settler-invader superiority positioned Aboriginal students as inherently inferior. This false assumption contaminated the objectives of residential schools and led to the systematic suppression of Aboriginal knowledge, languages, and cultures (1:251, 331–409). The report argues that this demeaning and ethnocentric attitude lingers in current policies that purport to work on behalf of Aboriginal peoples. Although this false assumption is no longer formally acknowledged, this does not lessen its influence on contemporary policies or mitigate its capacity to generate modern variants (1:249, 252–3). The report proposes that Canada dispense with all notions of assimilation and subordination, and develop a new relationship with Aboriginal peoples based on sharing, mutual recognition, respect, and responsibility.

Today, Indigenous peoples around the world continue to feel the tensions created by a Eurocentric educational system that has taught us not to trust our own tribal communities' knowledge, and by an increasingly fragile environmental base that requires us to rethink how we interact with the earth and with each other. We are also becoming increasingly aware of the limitations of technological knowledge, of the possibilities and potential of Indigenous knowledge, of the nature of our loss, and of the desperate need to repair our own systems. These tensions create many quandaries that touch on issues of diversity, inclusivity, and respect.

It is vital to protect Indigenous knowledge, not only for the sake of Indigenous peoples in their own environments, but also to raise general awareness about the vitality of Indigenous knowledge and its dynamic capacity to help solve contemporary problems. Scientists have begun to recognize its potential, but there is still much work to be done. In order to protect Indigenous knowledge, we must consider the Eurocentric biases and cultural appropriations that are endangering Indigenous peoples' cultures and languages, for these cultures and languages are the source of Indigenous knowledge. In schools, we must engage in a critique of the curriculum and examine the connections between and the framework of meanings behind what is being taught, who is being excluded, and who is benefiting from public education. We must centre Indigenous knowledge by removing the distorting lens of Eurocentrism so that we can immerse ourselves in systems of meaning that are different from those that have conditioned us. As we embrace this process, we can begin to untangle the knots our minds and practices have created in the web of knowledge so that we can weave a whole new cloth with threads that create a coherent but diversified pattern. Thomas Berry, author of *The Dream of the Earth*, suggests: "It's all a question of story. We are in trouble just now because we do not have a good story. We are in between stories. The old story, the account of how we fit into it, is no longer effective. Yet we have not learned the new story" (Berry, 1990, 123).

WHAT IS INDIGENOUS KNOWLEDGE?

Due to the exclusionary culture and curricula that have long permeated Eurocentric educational institutions, little is known about Indigenous knowledge. However, the framework and details of Indigenous knowledges are gradually emerging from Indigenous peoples' collective experiences and from the products of their minds and hearts. Knowledge of Indigenous peoples is embodied in dynamic languages that reflect the sounds of the specific ecosystems where they live and maintain continuous relationships. The multiplicity of prefixes and suffixes in Indigenous languages can thoroughly describe any experience, and thus, these languages are capable of describing scientific discoveries of everything that can be experienced. All Indigenous knowledge flows from the same source: the relationship of Indigenous peoples with the global flux, their kinship with other living creatures, the life energies as embodied in their environments, and their kinship with the spirit forces of the earth. Through time, Indigenous peoples have collectively developed a complex array of interrelated and inseparable relationships with the world around them. Their knowledge manifests itself in many linguistic and non-linguistic forms, such as stories, art, songs, prayers, ceremonies, and technologies, as well as in forms of the spirit, such as dreams and visions.

It is not helpful to attempt to define Indigenous knowledge, for it is a comparative knowledge system that should not be pounded into Eurocentric categories. Indigenous knowledge involves ecosystems and Indigenous peoples' relationships with them. Indigenous knowledge allows those who move within it to discover everything about the world they inhabit that they can discover with the tools of language and culture. To understand the meaning of life, Indigenous peoples need to re-establish relationships with the ecological order, for local ecological forces have always provided the most important lessons of Indigenous life. From their ecological sensibilities, Indigenous peoples derived their world views, languages, knowledge, order, and solidarity. It is essential to understand these ecological forces if we are to understand Indigenous contexts and thought.

Many definitions of Indigenous knowledge stress the principle of totality or holism. The RCAP has described Indigenous knowledge as "a cumulative body of knowledge and beliefs, handed down through generations by cultural transmission, about the relationship of living beings (including humans) with one another and their environment" (1996, 4:454). In English translations, the Inuit define their traditional knowledge as practical teaching and experience passed on from generation to generation. It is a total way of life that comprises a system of respect, sharing, and rules governing the use of resources. It is derived from knowing the land where they live, including knowledge of the interrelationships among everything in their environment. Inuit knowledge is rooted in the health, culture, and language of the people. It comes from the spirit and gives credibility to the Inuit. The Inuit assert that their holistic world view cannot be separated from the people who hold it. It is using the heart and head together in a good way. It is dynamic, cumulative, and stable. It is truth and reality (Environmental Assessment Workshop 1995, cited in Emery 1997).

In English translation for the Dene Cultural Institute, Emery (1997) writes:

Traditional environmental knowledge is a body of knowledge and beliefs transmitted through oral tradition and first-hand observation. It includes a system of classification, a set of empirical observations about the local environment, and a system of self-management that governs resource use. Ecological aspects are closely tied to social and spiritual aspects of the knowledge system. The quantity and quality of TEK [traditional ecological knowledge] varies among community members, depending on gender, age, social status, intellectual capability, and profession (hunter, spiritual leader, healer, etc.). With its roots firmly in the past, TEK is both cumulative and dynamic, building upon the experience of earlier generations and adapting to the new technological and socioeconomic changes of the present.

To Indigenous peoples, their lands, waters, and air count as interrelated parts of their knowing and of their identity. Indigenous people are their environment. This principle of totality is problematic for the Western mind. This is not how Western science works, and including macro concepts such as natural resources as part of knowledge makes it difficult for Western political institutions to legislate protection for it.

To speak about Indigenous knowledge in this manner presumes a uniformity among diversities. However, Indigenous knowledge is subject to many understandings and interpretations (Battiste and Henderson 2000), and no universal perspective on Indigenous knowledge exists. The diversity of Indigenous knowledge is a unifying idea among Indigenous peoples: each group comprises a unique blend of human variation, and their overarching humanity is manifested in their diverse attitudes, values, beliefs, and relationships.

All of humanity's knowledge about the natural realm is drawn from experience and perception, in other words, from sensations. Through language and other modes of inference, people transmit these sensations to others. Not all people value experience equally, nor do they put equal faith in their perceptions. Different languages create many different modes of inference about sensations and how to measure them. Science is all about how best to measure sensations. Scientific measurements then give rise to a theory about the nature of existence. Religion and philosophy are two other disciplines that serve the purpose of coming to know the world. Indigenous peoples have their own science or way of knowing the world, but it is a concept that embodies a way of life, an intimacy and directness with nature. The fragmented systems of Eurocentric knowledge have subdivided the knowledge of Indigenous peoples into separate disciplines, such as art, culture, religion, literature, science, medicine, and language. In such fragmentation, the embedded essences of Indigenous knowledge are lost in their visible manifestations.

It is difficult to find current literature about Indigenous knowledge that has not been tainted by Western perceptions of knowledge. Few studies are inclusive enough to embrace the complexities and dynamics of Indigenous knowledge. There are books available about different Indigenous cultures, and there is literature written by Indigenous authors, but one cannot fully come to know Indigenous knowledge by reading or by other vicarious experiences, especially when these experiences unfold in non-Indigenous languages. Books and vicarious experiences externalize Indigenous knowledge, but Indigenous knowledge is an intimate relationship. One comes to know only through extended conversations, enriched experiences, and awarenesses of particular ecologies. Elders humbly understand the depth of Indigenous knowledge when they reject being described as experts.

The disciplines of anthropology and ethno-botany have been developed using the knowledge of Indigenous peoples; however, this does not mean that the anthropologists and ethno-botanists who have studied Indigenous peoples can call themselves experts on Indigenous knowledge. The scientists have merely come to know some fragments of Indigenous knowledge. They have, in effect, only a partial sense of the knowledge of a group of people. This fragmentation of knowledge contributes to the Western scientific phenomenon of an observer-created reality. While it supports the structure of Western science, it does little to advance understanding of Indigenous knowledge.

Current literature about Indigenous knowledges does not address the issue of categorization. Ethno-scientists happily separate Indigenous knowledge from its media of songs, stories, kinship ties, and spiritual relationships with the environment. Breaking Indigenous knowledge down into separate categories serves particular purposes that Western scientists view as valuable. The holistic paradigm of Indigenous knowledge is ignored because the Indigenous facts are perceived only through the categories Western science has imposed on them. These categories impose limits on what counts as knowledge and what does not, yet no one can decide in the abstract whether a given classification is justified. The result is tension and uncertainty about the legitimate uses of and definitions of Indigenous knowledge.

Faced with the crisis of Eurocentric subjectivity, Indigenous peoples have turned to legal regimes to define the boundaries of their knowledge. Unfortunately, this exercise is also fraught with problems, for Western legal systems have never valued folk knowledge, and their legal rules are often too limiting to protect the principle of totality of Indigenous knowledge. Ultimately, it is up to Indigenous peoples to frame Indigenous knowledge, and it should be protected as a *sui generis* right by all legal systems (Smith 1997).

In 1993, Dr Erica Irene Daes, special rapporteur and chairperson of the Working Group on Indigenous Populations, reported to the United Nations on how best to protect the heritage of Indigenous peoples around the world. The Working Group on Indigenous Populations is now one of the largest attended gatherings in the United Nations pertaining to Indigenous peoples, and it is the site of the most comprehensive collection of paradigms of Indigenous knowledge.

Dr Daes has stated that Indigenous knowledge is not merely a collection of objects, stories, and ceremonies, but a complete knowledge system with its own languages, its own concepts of epistemology and philosophy, and its own scientific and logical validity. She has underscored the central role languages play in recording Indigenous peoples' heritage and in transmitting it from generation to generation. She has emphasized that legal reforms must recognize the unique and continuing links between Indigenous knowledge and the ecosystems, languages, and heritage of Indigenous peoples. She has also reported to the United Nations

Sub-Commission on Prevention of Discrimination and Protection of Minorities that such legal reforms are vital to a fair legal order, because Indigenous peoples cannot survive or exercise their fundamental human rights as distinct nations, societies, and peoples without the ability to conserve, revive, develop, and teach the wisdom they have inherited from their ancestors (Daes, 1993, 13). Dr Daes' statements are considered to be the best-articulated understanding of Indigenous knowledge.

In *Guidelines and Principles for the Protection of the Heritage of Indigenous Peoples* (1994, 6), the special rapporteur of the United Nations Economic and Social Council offered this definition of Indigenous heritage:

12. The heritage of Indigenous peoples includes ... all kinds of scientific, agricultural, technical and ecological knowledge, including cultigens, medicines and the phenotypes and genotypes of flora and fauna; human remains ...
13. Every element of an Indigenous peoples' heritage has traditional owners which may be the whole people, a particular family or clan, an association or society, or individuals who have been specially taught or initiated to be its custodians. The traditional owners of heritage must be determined in accordance with Indigenous peoples' own customs, laws and practices.

Many modes of Indigenous knowledge exist, and no single definition should be adopted universally (Brooke 1993). Indigenous knowledge has few internal boundaries, and attempting to subdivide Indigenous knowledge into separate Eurocentric legal categories such as culture, art, intelligence, science, and medicine, or into separate elements such as songs, stories, science, or sacred sites is inappropriate. Most Indigenous languages make no such distinctions. Elders have argued that all elements of Indigenous knowledge within a language group should be managed and protected as a single, interrelated, and integrated whole.

Indigenous knowledge can be conceptualized as the web of relationships between Indigenous people and the ecological world at a specific location, which is why elders and scholars are reluctant to discuss "universal" facts or physical "laws." Any framework for understanding or protecting particular perspectives of Indigenous knowledge must be contextual, decentralized, and respectful of the linguistic categories, rules, and relationships unique to each knowledge system. Furthermore, the modern issue of how best to conceptualize Indigenous knowledge must be handled with great sensitivity because of the history of Western appropriation of Indigenous knowledge in the past.

REFERENCES

Battiste, M. and J.Y. Henderson. 2000. *Protecting Indigenous knowledge and heritage: A global challenge.* Saskatoon, SK: Purich Publishing.

Berry, T. 1990. *The dream of the earth.* San Francisco: Sierra Club Books.

Blaut, J. 1993. *The colonizer's model of the world: Geographical diffusionism and Eurocentric history.* New York: Guilford Press.

Brooke, J. 1993. Gold miners and Indian Brazil Frontier War. *New York Times*, September 7, A4.

Daes, E. 1993. *Study on the protection of the cultural and intellectual property rights of Indigenous peoples.* E/CN.4/Sub. 2/1993/28. Sub-Commission on Prevention of Discrimination and Protection of Minorities, Commission on Human Rights, United Nations Economic and Social Council.

Emery, A.R. 1997. *Guidelines for environmental assessments and traditional knowledge.* Unpublished. Ottawa: Centre for Traditional Knowledge.

Memmi, A. 1965. *The colonizer and the colonized.* (Howard Greenfield, trans.) New York: Orion Press.

– 1969. *Dominated man: Notes toward a portrait.* Boston: Beacon Press.

Minnich, E. 1990. *Transforming knowledge.* Philadelphia: Temple University Press.

Phillipson, R. 1992. *Linguistic imperialism.* Oxford University Press.

Royal Commission on Aboriginal Peoples (RCAP). 1996. *Report of the Royal Commission on Aboriginal Peoples.* 5 vols. Ottawa: Canada Communication Group.

Smith, G. 1997. *Kaupapa Maori theory and praxis.* Ph.D. dissertation. University of Auckland, Auckland, New Zealand.

9

Colonialism and Capitalism: Continuities and Variations in Strategies of Domination and Oppression

VANAJA DHRUVARAJAN

HOW I BECAME INTERESTED IN THE STUDY OF GLOBALIZATION

My entry into North America as a graduate student in the '60s made me acutely aware of the impact of the colonization of India on the devaluation of us as a people and on the devaluation of our culture. Such devaluation had become a part of legitimate academic common-sense. My struggles against such a state of affairs as a student was feeble at best since questioning the judgment of learned professors would have sounded a death knell to my career as a student. Experiences of marginalization and devaluation of my culture and race, because of my colonial background, were particularly difficult burdens to have to carry after having had to endure devaluation as a woman in a patriarchal Hindu family in patriarchal India. I was surprised to see that patriarchy in addition to ideologies of colonization and imperialism, were alive and well in a land that boasts of inalienable rights of the individual, democracy and freedom for all. Joining the faculty of a Canadian university did not improve that situation much and my life as an academic has been one of relentless struggle to belong, to be accepted and respected. This struggle in a significant sense has led me to choose research topics in sites where race, class, and gender intersect.

My fond hope that colonized countries after political emancipation could look forward to becoming economically prosperous and self-reliant proved to be unrealistic. What was in fact happening was that new strategies were being developed to control and direct these previously colonized countries by the developed countries of the West. Colonization was justified as being for the good of the colonized. It was argued that it was the White Man's burden to civilize the uncivilized world. But such arguments proved to be untenable in view of many movements in these colonized countries to get rid of the colonizers. It had become evident that direct control and domination is not possible; therefore new ways and means of controlling and dominating the Third World were being contemplated. The strategy was the imposition of a neoliberal Western model of development on these countries. It was again interpreted as the best thing for these countries because Western experts know best. Thus the old oppressions were taking new forms.

NEGATIVE CONSEQUENCES OF GLOBALIZATION

The devastating impact of these neocolonial policies is evident everywhere, but is particularly pronounced among historically marginalized peoples because the new policies are superimposed upon existing structures of domination and subordination. Thus developing countries are vulnerable because of their histories of colonization and imperialism (Bello 1996).[1] With the implementation of these policies, the flow of wealth from the developing countries to Transnational Corporations (TNCs) located in the North has increased to such an extent that Susan George refers to this exploitation as the greatest highjack in history carried out with legal impunity (George 1999). The experts' faith in the ideology of neoliberalism and commitment to it are deep and far-reaching. To spread this ideology, many strategies have been devised. Most important of these is the establishment of research centres and think tanks. Books and articles have been written. And in North American universities, students are carefully trained in the tenets of this paradigm. Among these students, many are from developing countries, recruited to implement the corporate-sponsored globalization project. They are guided by a neoliberal ideology that, it is argued, provides the engines of growth (Norberg-Hodge 1996).

Women everywhere have taken the brunt of these changes because of patriarchal structures. Women in developing countries suffer most, since they not only are victims of the "gender-gap" but also of the "development-gap" (Dhruvarajan and Vickers 2002; Marchand and Runyan 2000).

As Weichterich writes, "In the fierce undercutting that pits one country against another, 'globalized woman' is burnt up as a natural fuel; she is the piece-rate worker in export industries, the worker living abroad who sends back foreign currency, the prostitute or catalogue bride on the international body and marriage markets, and the voluntary worker who helps to absorb the shock of social cutbacks and structural adjustment. The strategic function of 'globalized woman', within the broader project of globalization driven by economics and politics, is the execution of unpaid and underpaid labour " (Weichterich 2000, 167). Environmental depletion and degradation have been an integral part of this project of globalization. As Shiva (1994: 3) points out, environmental problems lead to health problems among people. There are subtle and complex interconnections among the diseases of the human body, the decay of ecosystems, and the breakdown of civil society. In spite of widespread devastation and suffering caused by these policies, the sponsors of corporate globalization continue to implement it with a zeal reminiscent of religious fanaticism. As McMurtry (1998) writes, the underlying principles of this project are life-blind, without any concern for morality and ethics.

DISCOURSE OF INEVITABILITY AND PROMOTION OF CORPORATE SELF-INTEREST

This neoliberal project of the corporate elite is proposed as the inevitable way to economic development. It is also argued that it has naturally evolved and therefore humanity has no option but to accept the triumph of capitalism and try to cope the best we can (Teeple 2000). These arguments are made in spite of the evidence that the neoliberal agenda is a carefully crafted and executed plan of action (Chomsky 1997). When we consider the various deliberations of the corporate elite shrouded in secrecy, the claims of inevitability and the natural evolution of corporate-sponsored globalization have to be rejected. Elites have been able to sustain this stance because the media is under their control. Managing public opinion to accept their dictum as legitimate has been the preoccupation of corporate-controlled media.[2]

The argument that globalization is the only way and that a new world order has to be ushered in makes sense when we consider these arguments from the perspective of the corporate elite. For the corporate elite, this is the only way; otherwise their agenda of profit maximization cannot be achieved (Mies and Shiva 1993). A very disturbing revelation about the state of affairs is that the world-view of the corporate elite is considered as the only legitimate one. Their interests are considered paramount. In fact, corporate interests are interpreted as people's interests. Even democracy and freedom is passed through the corporate lens.[3] To be free is to have free trade and to have democracy means unfettered conduct for corporations. Catering to the needs of corporations has to be the national agenda of democratically elected governments in spite of public opinion against such measures (Nader and Wallach 1996).[4]

The mandate of capitalism is considered so crucial that it almost borders on the sacred. When we examine treatises of ideologues of capitalism such as Friedman (2000), this point become clear. He considers implementation of the project of globalization as nothing short of exemplary American patriotism. The unquestioned superiority of America's leadership is hailed as unassailable. He exhorts everyone to join in and spread Americana unfettered by any moral principles.

TRADE AGREEMENTS AND DEVELOPING COUNTRIES

Trade Related Intellectual Property Rights (TRIPS) agreements are a key concern of developing countries, which are in danger of losing much of the wealth produced over centuries by multitudes of people to corporate greed (Shiva 1997; Shiva and Holla-Bhar 1996). These agreements need to be changed to outlaw biopiracy and patents on life. According to the trade agreements negotiated by the World Trade Organization (WTO), there are many conditions the developed countries are expected to honour in their dealings with developing countries. They include providing better access to them in their markets; providing technology transfer; stopping the practice of dumping subsidized products (particularly in agriculture); and increasing investment to trigger growth in the 'developing' economy. But none of these conditions are ensured to the satisfaction of developing countries even though the TNCs from the developed countries have obtained all the advantages and have profited from the misery of the people in developing countries. There have been many demonstrations against such corporate-sponsored globalization across the globe and a plea for the TNCs to reconsider their strategies.[5]

There are widespread feelings of frustration and a sense of powerlessness in Third World countries. Clear awareness of being marginalized and manipulated has resulted in a sense of alienation.[6] Strategies of arm-twisting, blackmail, and intimidation from big trading powers have silenced and forced acquiescence. Thus a "consensus without consensus"[7] has been obtained in the meetings of organizations such as the WTO.[8]

PROSPECTS FOR THE FUTURE

The situation looks bleak to a majority of people around the world as corporate-sponsored globalization sweeps the globe and the powers of corporations become entrenched. Corporate rights have become sacred rights that override even the rights of sovereign states to protect their citizens.

But there are some silver linings around the dark clouds of neoliberalism enveloping the globe. Global economic institutions are losing credibility in their promise to bring about economic prosperity for all. No matter how strong the institutions appear to be, these feelings of discontent continue to spread. The hegemony is beginning to unravel (Gramsci 1987; George 1999; Bello 2000). There is growing awareness of the duplicity of these institutions' dealings and their manipulative tactics to serve the interests of TNCs. The heads of institutions such as WTO are not democratically elected but are bureaucratic appointments, and *their* goal is to implement a neoliberal agenda and not to serve the interests of people in general. The dictum that what is in the interest of corporations is in the interests of all is being questioned. There is accumulated evidence to demonstrate that bad theory hurts the welfare of people and even kills them (Hahnel 1999); theories of development derived from the neoliberal paradigm are indeed doing just that (Chossudovsky 1997; Korten 1996; McMurtry 1998). There is a growing realization that the negative impact of corporate globalization is intrinsic, not an accidental outcome (Mies and Shiva 1993; Korten 1996). The number of people demonstrating against globalizing institutions every time they hold a meeting shows a growing unrest among people.[9]

The question for most people around the world is not whether they are "for" or "against" trade and technology. The concern is about the terms of this trade and the imposition of different types of technology. The issue is one of democracy and freedom for all people, not just for the elite (Langdon 1999).[10]

The problem of domination and exploitation is not new for people in the South because under colonial rule, the citizens of colonies did not have any democratic rights. This arrangement was rationalized on the basis of the assumed superiority of Western cultures and their mission to civilize the heathen. Since the dissolution of colonial rule, there has been some recognition of the rights of all people to practise their culture. But this politics of recognition has resulted in liberal indifference at best. This position is articulated succinctly by Charles Taylor (1994), who states that Westerners should recognize that there are many different cultures and that all peoples should be allowed to practise their cultures. But he argues that no statements about the "validity" of other cultures can be made since Westerners do not know them. He does not argue for engaging in a discussion to understand and appreciate these different cultures, but rather prefers to leave them alone. Such liberal indifference is not conducive to the development of respect and acceptance among members of different cultures. More important, indifference does not challenge the assumption of the superiority of Western culture over other cultures.[11] Thus it does not provide conditions conducive for democracy to thrive across the globe. Instead it creates conditions within which Westerners can continue to impose their way of life under the guise of expertise in the missionary tradition of offering a choice of a better life to people while in fact pursuing their own self-interest.

ALTERNATIVES

The charge that critiques of corporate globalization have not come up with clear alternatives is not justified. They have not come up with one grandiose design because that is exactly what is wrong with the corporate globalization project. Being sensitive to variations in aspirations of people depending on their culture and history, and building enabling structures and ideologies to promote multiple ways of life is what is being advocated by critics. Corporate globalization is literally smothering all such efforts and thereby scuttling all other projects in their infancy.

There is an emerging consensus among the critics of corporate globalization to provide alternatives to the neoliberal paradigm. The goal is to promote conditions for the adaptation of diverse ways of life including diverse economic paradigms. The crucial point to recognize is that people should be free to choose a particular way of life that is in keeping with their history and culture. They should be free to choose appropriate terms of trade, appropriate technology, and an appropriate form of government that works best for them. The intent is to empower people and communities so that they can participate in decision-making processes that determine their ways of life. The argument is for decentralized power structures rather than centralized bureaucracies. The call is made to build enabling structures to nurture and promote diverse ways of life.[12] With such consensus regarding goals, several alternative methods for achieving those goals are proposed. Space does not permit an exhaustive discussion of such models. Here I discuss four to illustrate their arguments.

One paradigm is suggested by Amartya Sen (2000), a Nobel prize winner for economics. In his book, *Development as Freedom*, he takes a holistic approach, emphasizing that the purpose of development is the enhancement of human capabilities. Economic production is not an end in itself but is one of the means (albeit an important one) to enhance human capabilities. People are not just recipients of development policies. They are also agents who participate actively in the process. Such participation is facilitated in contexts where their capabilities are enhanced. This requires that conditions that are conducive to freedom and autonomy, provide opportunities for all people to increase their potential, and nurture their health and well being, prevail. All development policies should reflect this orientation. In concrete terms, this translates into making the health and well-being of people and environmental preservation an integral part of development policies. Thus Professor Sen has a strong conviction that by integrating ethics and morality into its framework, capitalism can accomplish the goal of empowerment of all people and ensure their well-being. He seems to envisage the utopian dream of Adam Smith and Friedrich von Hayek of a humanity united by peacefully competing enterprises.

Showing that corporate globalization is essentially the same as imperialism (since its goal is to further enrich and maintain the privilege of those who always had both riches and privilege), scholars such as Petras and Velmeyer (2001) and Ugarteche (1997) argue for strong nation-states to transcend class conflict to ensure the well-being of all. They do not see a conflict between a strong state and the private market. They show how industrial development in the developed countries took place over time through state protection and the nurturing of domestic industries. They do not agree that export-led growth is the way to achieve prosperity. Instead they argue for the development of a strong internal market, technology and infrastructure, before integration into the global market. A strong interventionist state, they argue, can successfully negotiate with the market to produce economies capable of dealing with external markets from a position of strength. Foreign investment should only be undertaken with agreements that are fair and just. It is important that capital be rooted in the community and not transferred at the whim of speculators, thereby destabilizing the economy, and consequently social life in general. They argue that the neoliberal strategy of integrating weak underdeveloped economies of the South with well-developed strong economies of the North means the former will be integrated as dependent economies, destined to remain that way. The end result will be the granting of a monopoly over power and privilege to the corporate elite in perpetuity. These authors do not seem to be concerned about the growth of unmanageable state bureaucracies.

Scholars concerned with the environment argue for decentralized economies with vibrant communities (Mander and Goldsmith 1996). They argue for the establishment of enabling institutional structures to sustain and promote such economies and communities. These scholars are convinced that a democratic way of life thrives through the empowerment of people and communities. They argue that protecting the environment and nurturing diverse cultures and ways of life in cooperation and mutual respect would best be achieved under such circumstances (Kumar 1996). They further encourage living in harmony with nature and accepting that the assumption of infinite growth in a finite earth is unsustainable, in order to ensure a positive legacy for future generations. In Walden Bello's words, "We are talking ... about a strategy that consciously subordinates the logic of the market, the pursuit of cost efficiency to the values of security, equity, and social solidarity. We are speaking in short, about re-embedding the economy in society, rather than having society driven by the economy" (2000, 11). From a sustainability perspective, producing everything that is needed locally and depending on other countries for only those products that cannot be produced at reasonable cost should be the principle to follow. According to Mander (1996), fair trade in goods that are produced cheaply and in excess of local consumption is in the best interests of people in local communities. Self-sufficiency in food is a must; otherwise it can be used as a weapon to impose controls as it is now being done by the TNCs. Engaging in economic activities in such a way that they do not deplete and pollute the environment is considered vital in this model.

Finally, emergent feminisms offer models that are based on a way of life emphasizing harmony with nature, cooperation and mutual respect among people, and a pervasive concern for justice and care of all people (Dhruvarajan 2002). These visions promote a holistic interdependent world-view and social relations that emphasize cooperation (Lorde 1984; Mies and Shiva 1993). They also recognize the finite nature of earth's resources and our own mortality and insist that we take these factors into account in organizing our lives (Gross 1996). Thus the values of democracy and the human values of caring, sharing, and a sense of community provide the guiding principles to organize social and economic life. Promoting human dignity and welfare is front and centre in all these paradigms. The privileging of economic aspects of life and subordination of all other aspects as the neoliberal paradigm does is seen as dehumanizing to all people, including those who gain financially. In addition to destroying the environment, such privileging of economy deprives future generations of their rightful legacy. Providing for the needs of all people and living a simple but meaningful life is considered important. The call is for grassroots involvement in determining the direction of social life by strengthening civil society. Instead of segregating private and public life, these feminists advocate integration of economic and non-economic spheres with guidelines for behaviour that promote well being of all people. They insist on resolving conflicts through peaceful means and avoiding diversion of resources towards non-productive military build-ups.

CONCLUSION

Confronting and resisting the entrenched power of global capitalism is daunting indeed. But all those who are concerned about the well-being of humanity cannot afford to turn away because of the monumental nature of this struggle. It is up to intellectual and moral leaders to theorize and explain the severity of the problem, articulate visions for the future, make connections between local and global concerns and raise critical awareness among all people. Educational institutions have a crucial role to play in informing and educating students regarding the negative impact of globalization. Building alternative media to create a well-informed community of citizens regarding the impact of corporate globalization is necessary to support counter-hegemonic movements.[13] In the final analysis, it is up to people at the grassroots level who have the wisdom that comes from the experience of daily struggles to provide practical solutions to these problems. The solutions necessarily have to be diverse depending upon the history and culture of a given locality. To come up with and implement these solutions, people have to work together. Coalition building across differences is crucial in this context. Resistance bears fruit only when there is unity of purpose and solidarity in struggles. Because people come from diverse backgrounds, their interests may not always coincide; sometimes they might even appear to be antagonistic. It is sometimes necessary to distinguish between strategic and practical interests by identifying long-term and short-term goals.[14] As Michael Albert argues, "We must raise social costs to those elites to the point where they decide that giving in is their best course of action ... Elites don't respond to reason or to morality. They respond to movements that will do more damage to their interests if elites don't give in than if they do" (2000, 5).

It is important for all of us to move beyond liberal indifference and make a sincere effort to know one another's ways of life. When we interact as equals to explore our experiences, we often discover that we have many things in common along with our differences. Exploring these issues requires different ways of thinking and acting. It is only when we engage in such dialogue that we are able to develop solidarity across differences. Such solidarity is imperative to achieve the goal of dismantling the powers of the corporate globalizers and to reclaim our right to choose to live a way of life that is meaningful to us. It is only under such conditions that we can construct a world that is just and caring for all.

NOTES

1 Free market rules and regulations are imposed on poor, helpless countries but the rich and powerful countries have always had help for their industries, which were nurtured to grow. The same is true with regard to rich and poor people within capitalist democracies – children of the rich are supported while the poor have to fend for themselves.
2 The trend of creating media monopolies means that only elites can dictate what the public will know. The corporate makeover of the news becomes effective when diverse interpretations of news are scuttled in favour of a few corporate-managed news releases.

3 Petras and Veltmeyer state that "As democracy has been redefined as centralized elite decision-making with elections, the role of citizens as protagonists of public policy debates has declined. The result is greater voter apathy, increased abstention, rejection of political incumbents, 'anti-voting' and increased resorting to extra-parliamentary action" (2000, 71).

4 In the 1980s Canada's public opinion was solidly against free trade. In spite of this, both conservative and liberal governments signed the free trade agreements.

5 Countries from the South are working together to achieve this objective. In addition there have been widespread demonstrations by farmers, for example in India. There they demanded that either India negotiate to keep food out of the agreement or that India get out of the WTO.

6 There have been demonstrations every time the globalizing economic organizations have conducted meetings. All discussions take place in secrecy even though the agreements reached have far-reaching consequences for everyone. The demonstrations in Seattle, Quebec City, Genoa, and Cancun are only a few examples.

7 Chomsky (1997) argues that manipulating attitudes and beliefs of the general public is routine since in a democracy citizens cannot be dictated to do the bidding of those who are in charge of running the affairs in society, namely the elites. Manufacturing public opinions to promote corporate interests is a big industry in North America. The so-called Third way advocated by people like Tony Blair and Bill Clinton is nothing but a public relations exercise to make the public swallow the bitter pill of corporate globalization.

8 The September 11 tragedy is being used by the big trading powers to push their own agenda. How to consolidate and entrench the North's hegemony everywhere is being discussed. Deliberate efforts are evident at every level involving enactment of laws to neutralize dissent against corporate domination and exploitation. After September 11, the problems for the countries of the South have worsened. Renewed efforts have to be made to address these issues. As the Focus on the Global South writes, "We must link our existing and common demands on neoliberal globalization to an agenda that includes a clear voice against militarization and imperialism and proclaiming peace, cultural and religious freedom and self-determination. This will be extremely difficult in a climate where *all* forms of dissent will be subject to much greater scrutiny and repression, and in a climate of heightened xenophobia and militarization. The cry from the establishment will be if you're not with us, you're against us" (http://*www.focusweb.org* Accesssed November 23, 2001).

9 Demonstrations against the corporate agenda and their secret dealings are becoming more and more vocal. People from many different backgrounds are joining in and trying to work out a cohesive plan of action. A number of critiques are being written and posted on the web and teach-ins are held every time these meetings take place to make people aware of the implications.

10 Six basic issues stressed are the following: poverty reduction; gender equality; environmental protection; people's participation in shaping their future; safeguarding rights of community based movements; and having economic and political leverage to establish national priorities.

11 Audre Lorde (1984) and Philomena Essed (1991) argue that Western culture does not allow for treating difference as normal and natural. Western culture is only capable of dealing with difference through the creation of a hierarchy of superiors and inferiors. Unless there is a fundamental paradigm shift, the pattern of behaviour in the West that has resulted in so much pain and suffering all through history (and continues to do so) will not change (Gray 1998).

12 The establishment of democratic structures of global governance under the auspices of the United Nations, which would support and nurture all people, is one option. In addition, we must build structures to empower people and communities around the globe. Such conditions challenge globalization to take place from the bottom-up rather than the top-down as it does now.

13 It is necessary to strengthen and nurture alternative media. At the present time there are valuable efforts, such as the *New Internationalist* magazine and *The Canadian Center for Policy Alternatives*. There are also websites such as, http://www.newint.org; http://www.attac.org; http://www.agp.org; http://www.ruckus.org; http://www.aseed.net; http://www.canadians.org; http://www.50years.org; http://www.zmag.org; www.worlbankboycott.org; www.whirledbank.org; www.wtowatch.org; www.corpwatch.org; www.WeAre Everywhere.org; to mention a few (All accessed November 23, 2001).

14 People from different parts of the world can help each other by providing support to one another in their struggles and by providing insight and knowledge about various strategies that work to confront and resist corporate globalization. For example, as Hahnel writes, "people in developed countries can help those in developing countries: Many third world unions and grassroots organizations appreciate help from first world progressives in their campaigns ... to publicize abuses – particularly when our multinational corporations are the perpetrators ... Occasionally, when their struggle is at a crucial stage, third world movements for human, political, and labour rights ask us to pressure our governments and/or international

organizations to take up economic sanctions, as was the case in the struggle against apartheid in South Africa and is now the case in the struggle for democracy in Burma" (*http://www.zmag.org* Accessed November 23, 2001). And people in the Third World can help those in the First by showing how participatory democracy works as in Porto Alegre as discussed by Judy Rebick (*http://www.zmag.org* Accessed November 23. 2001). The struggles of farmers in India against bio-technology and patenting of plants and other life forms can provide insights to develop strategies of resistance in the North.

REFERENCES

Albert, M. 2000. The movements against neoliberal globalization from Seattle to Porto Alegre. Talk delivered electronically in the Anti-globalization Conference in Athens in March.

Bello, W. 1996. Structural adjustment programs: 'Success' for whom? In *The case against the global* Albert *economy*, ed. J.M. and E.G. Goldsmith. San Francisco: Sierra Club Books.

– 2000. From Melbourne to Prague; the struggle for a deglobalized world. Talk delivered at a series of engagements on the occasion of demonstrations against the World Economic Forum in Melbourne, Australia, 6–10 September.

Chomsky, N. 1997. *Perspectives on power: Reflections on human nature and the social order.* London: Black Rose Books.

Chossudovsky, M. 1997. *The globalisation of poverty.* London: Zed Books.

Essed, P. 1991. *Understanding everyday racism: An interdisciplinary theory.* London: Sage.

Dhruvarajan, V. and J. Vickers. 2002. *Gender, race and nation: A global perspective.* Toronto: University of Toronto Press.

Dhruvarajan, V. 2002. Feminism and social transformation. In *Gender, race and nation: A global perspective,* ed. V. Dhruvarajan and J. Vickers. Toronto: University of Toronto Press.

Friedman, T.L. 2000. *The lexus and the olive tree.* New York: Anchor Books.

George, S. 1999. A short history of neo-liberalism: Twenty years of elite economics and emerging opportunities for structural change. Talk delivered during the Conference on Economic Sovereignty in a Globalising world, in Bangkok, in March. (*http://www.millennium-round.org* Accessed November 23, 2001)

Gramsci, A. 1987. Class, culture and hegemony. In *Culture, ideology and social process,* ed. T. Bennett et al. London: Open University Press.

Gray, J. 1998. *False down.* New York: New Press.

Gross, R.N. 1996. *Feminism and religion.* Boston: Beacon Press.

Hahnel, R. 1999. *Panic rules.* Cambridge, MA: South End Press.

– 2001. *Imperialism, human rights and protectionism: A question and answer session on globalization.* http://www.zmag.org Accessed November 23, 2001.

hooks, b. 1994. *Outlaw cultures: Resisting representations.* London: Routledge.

Korten, D.C. 1996. Global economy and the third world. In *The case against the global economy: And for a turn toward the local,* ed. J. Mander and E. Goldsmith. San Francisco: Sierra Club Books.

Kumar, S. 1996. Gandhi's swadeshi: The economics of permanence. In *The case against the global economy: And for a turn toward the local,* ed. J. Mander and E. Goldsmith. San Francisco: Sierra Club Books.

Langdon, S. 1999. *Global poverty, democracy and north-south change.* Toronto: Garamond Press.

Lorde, A. 1984. *Sister outsider.* Freedom Calif: The Freedom Press.

Mander, J. 1996. Facing the rising tide. In *The case against the global economy: And for a turn toward the local,* ed. J. Mander and E. Goldsmith. San Francisco: Sierra Club Books.

Mander, J. and E. Goldsmith (eds.) 1996. *The Case against the global economy: And for a turn toward the local,* San Francisco: Sierra Club Books.

Marchand, M.H. and A.S. Runyan. 2000. *Gender and global restructuring: Sightings, sites and resistances.* London: Routledge.

McMurtry, J. 1998. *Unequal freedoms: The global market as an ethical system.* Toronto: Garamond Press.

Mies, Maria and Vandana Shiva. 1993. *Ecofeminism.* London: Zed Books.

Nader, R. and L. Wallach. 1996. GATT, NAFTA, and the subversion of the democratic process. In *The case against the global economy: And for a turn toward the local,* ed. J. Mander and E. Goldsmith. San Francisco: Sierra Club Books.

Norberg-Hodge. 1996. Shifting direction: From global dependence to local interdependence. In *The case against the global economy: And for a turn toward the local,* ed. J. Mander and E. Goldsmith. San Francisco: Sierra Club Books.

Rebick, J. 2001. *Port Alegre* http://www.zmag.org Accessed November 23, 2001.

Sen, A. 2000. *Development as freedom.* New York: Anchor Books.

Petras, J. and H. Veltmeyer. 2001. *Globalization unmasked.* Halifax: Fernwood Publishing.

Shiva, V. ed. 1994. *Close to home: Women reconnect ecology, health and development worldwide.* Philadelphia, PA: New Society Publishers.

Shiva, V. and R. Holla-Bhar. 1996. Piracy by patent: The case of the Neem tree. In *The case against the global economy: And for a turn toward the local,* ed. J. Mander and E. Goldsmith. San Francisco: Sierra Club Books.

Taylor, C. 1994. The politics of recognition. In *Multiculturalism: Examining the politics of recognition,* ed. A. Gutmann. New Jersey: Princeton University Press.

Teeple, G. 2000. *Globalization and the decline of social reform*. Aurora, Ontario: Garamond Press.

Ugarteche, O. 1997. *The false dilemma: Globalization–opportunity or threat?* New York: St Martin's Press Inc.

Weichterich, C. 2000. *The globalized woman: Reports from a future of inequality*. New York: Zed Books.

10

The Brave New World of Professional Education

LINDA MUZZIN

"Never paint what's wrong with a sugar-coat"[1]

What I am about to say about the university is not what people want to hear. At first, I did not want to pay attention to academic capitalism, and what it signified about the university. I blamed the unethical behaviours I saw on the individual professors and administrators who engaged in them. Later, when the systemic nature of the problem became apparent, I pretended that only *part* of the university was affected by academic capitalism – the pharmaceutical sciences. I thought that I would be able to go elsewhere inside the academy to teach about emancipation, which I thought was the purpose of the university. In fact I have taught a course called *Nurturing Professional Education* in the Higher Education graduate program at OISE/UT which dwells on the positive achievements of professions such as nursing and social work in the community. We take the heartwarming approach of reading the visionary writing of educators who both critique and call for change in education. However, I am often disappointed that I cannot convey how bad things really are in terms of equity. And I firmly believe that if we cannot describe how bad things really are, we cannot escape complicity and truly work for equity.

Personally, I had a lot of trouble even coming to the point of being able to reflect on the problem. But in 1995, after a few years as a professor in the largest university-based pharmaceutical sciences program in Canada, I abruptly left the field. A sabbatical allowed me to travel across the country and express what I'd experienced to a tape recorder. Remembering was so gut-wrenching that I stopped regularly to disengage by communing with the beautiful landscapes. Particularly healing were the wide skies of the prairies and the cuestas north of Lake Superior. I also tricked myself into talking about what I had experienced in the form of tape-recording conversations with Peggy Tripp. She had similar memories of forestry science, and so was able to join in the tune I was singing – she knew the words. Finally, I interviewed hundreds of professors in the university-based professions, and heard many saying the same things I had at first been unwilling to even think. That was when I realized that the university itself was the problem, not the solution, and that perhaps by teaching in a professional faculty, I was part of the problem too.

The next challenge has been to *publish* the story of academic capitalism. There are very few venues willing to legitimize accounts that name the university – via corporate control of professional curricula and research agendas, knowledge production, and validation – as a key player in neo-colonialism. To take only one example, my proposed paper for a higher education conference on globalization in Vienna one summer was rejected with no explanation, although my colleagues assure me that its organizers usually accept everything that is submitted to it. Indeed, part of the paper you are reading was rejected by an editor of *The Canadian Review of Sociology and Anthropology* who had called for papers on the university in the twenty-first century. She explained very briefly, after my having to ask, that she wanted to spotlight important issues that cut across the university and did not just deal with one discipline. Fortunately, some of my analysis has appeared in two good collections by critical scholars this past year. I can now locate my work in a larger body of critical scholarship, which I review here very briefly first.

HIGHER EDUCATION FOR EMANCIPATION OR FOR COLONIZATION?

As even politically conservative texts such as McNeill's *An Environmental History of the Twentieth-Century World* (2001) make clear, our world is in trouble. McNeill's text, which won the 2001 World History Association Book Award, demonstrates repeatedly that human activities in the past century have upset what was a delicate ecological balance on earth. Occasionally, the text points to the positive results of community activism in decreasing air, soil, and water pollution and destruction of other life, while at the same time delivering the disquieting message that this kind of destruction cannot continue if life on earth is to survive. There is enough detail in this book, if one is familiar with the history of class oppression, gender discrimination, and colonialism from the 1500s on, to see the intensifying ecological damage done by the domination of a few nation-states over their own populations and much of the rest of the world during that period. There are hints that Aboriginal knowledges of agriculture left but a light footprint on the earth and that women and the poor initiated environmental cleanups after the excesses of both capitalism and communism. But like many books dealing with "just the facts," McNeill's cries out for an incisive critique of twentieth-century Western science and technology that has been forced on Indigenous and powerless populations everywhere and has globalized the subjugation of nature.

Such critiques are loud and clear in the work of feminist and Third-World environmental scholars. Perhaps the most cited of these is Vandana Shiva (1988, 1995, 1997), who has argued that White, androcentric technology – the pride of our dominant culture – is at the heart of the problem. As she explains, the White monocultural "technology transfer" that is occurring as part of globalization is a new form of colonization of the third world, this time based on professional knowledge. In McNeill's account, scientists seem to mitigate the excesses of politicians. What he fails to mention is that the professionals who work for the multinational corporations that are cutting down the Amazonian and Canadian rainforests are the same ones who were taught principles of "forestry science." Professionals who work towards patenting the rare genes of Indigenous peoples as well as their traditional agricultural products for pharmaceutical firms and agribusiness are trained in reductionist biotechnological science. Shiva argues that at the same time that White capitalist and patriarchal science and biotechnology is exported to the rest of the world, the activities of the transnationals promoting this knowledge are engaged in a dangerous reduction of the biodiversity of life. Without diversity, she argues, life is likely to eventually collapse.

Canadian scientist Rosalie Bertell (2001) has added the story of how White military research has been implicated in the production of large-scale radioactive contamination, toxic waste dumps, and ozone depletion. She argues convincingly that much of the scientific research that engages graduate students is directly or indirectly of interest to US national security, and that young professionals are unwittingly drawn into this system which threatens both social justice and the environment. Canadian activist and retired engineering professor Ursula Franklin similarly problematizes our unwitting participation in war-making; she argues that we willingly or unwillingly make a *financial* commitment to war every time we pay our income taxes (1999 [1990]: 77).

Fortunately, there are global peace, environmental, and social justice resistance movements against both war-making and transnationals, as well as reassertions of Indigenous rights. In the academy, perhaps the richest source of evidence about the patriarchal and capitalistic nature of scientific and professional knowledge can be found in the burgeoning White feminist literature critiquing science. Classic works in this tradition have brilliantly shown how mainstream biology has constructed women and sexuality as cultural stereotypes (e.g. Haraway 1978, Keller 1996, Martin 1991). Leading American theorists such as Harding (1986, 1998), Hubbard (1995; Birke and Hubbard 1995), and Haraway (1991, 1997) have transformed the sociology of science through their considerations of what a "postcolonial science" might look like. In Canada, feminist critics of science have problematized both science and technology, in particular, new reproductive technologies; science policy; the methodology of science; the animal-industrial complex; and most recently, new invasive biotechnological methods such as gene splicing, cloning, and xenotransplantation (Clark 2001; Burt and Code 1997; Franklin 1999; Noske 1997; Tokar 2001). In addition to the publication of classic historical works on academic capitalism (e.g. Noble 1977), there is an exponential growth of radical literature critiquing government, business, and academic complicity in environmental degradation and the globalization of oppression (Gaffield and Gaffield 1995; Marchak 1983, 1995; Tokar 2001; Tudiver 1999; Turk 2000). Most important, systematic coverage and critique of the activities of transnationals is delivered through listserves that summarize and problematize media coverage of education issues. Daily, critical scholars use emancipatory texts in their teaching and participate in the construction of emancipatory and environmental policy. At an unveiling of the "New Economy" initiative by the federal government's social sciences and humanities research funding agency, we were promised "exciting new possibilities." Scholars who knew better expressed skepticism.

Thus as intellectuals "in society," a sizable group of critical scholars in Canada are both present and engaged. However, little or none of this activity has penetrated into our professional schools, which are erecting buildings with their new-found capital, expanding their student numbers, raising their tuition, and redesigning curricula to include courses on the new technologies. How has this happened? How can our professional schools and our university administrators be so out of touch that they champion such neocolonial technology transfer? How can they fail to see that technology transfer is part and parcel of the growing gap between the rich and the poor as well as the degradation of the environment? What is holding this system in place?

A brief history of professional education, academic capitalism and the dismal record of universities on matters of equity

There is good documentation of the origins of dominant professions such as medicine and engineering in White middle-class Western patriarchal capitalism. In the twentieth century, biomedical and engineering solutions to social problems were sought out by capitalist nation-states (Noble 1977; Brown 1979) who gave these professionals the right to self-educate and, in terms of practice, self-regulate. From within, their reductionist technical approaches were occasionally and belatedly critiqued by scholars who called for more holistic and practice-based epistemology and pedagogy (e.g. Schön 1983, 1987) or a more client- or patient-centred approach (e.g. Bevis and Watson 1981). But such calls for accountablility to the masses have been largely unheeded. At best, activism for community responsiveness has transformed some biomedical professional curricula into contradictory mixes of what has been called "caring vs. curing" (Good and Good 1993; Bloom 1988). The academic environmental movement, including Canadian research on environmental racism and sexism (Thompson 2002; Barlow and May 2000) and the voices of First Nations peoples and women of colour have called for more attention to the environmental consequences of large-scale engineering interventions. Though architecture has responded faintly, engineering curricula have been largely resistant to these pressures, with ideas about sustainability squeezed between or inside largely technical courses or talked about as ideas for future curriculum renewal (Middleton 2000).

Part of the activist ineffectiveness in transforming technical curricula is due to the huge investment of capitalists and capitalist states in technical professional education that was sold to the public as a driving force of national economies, just as globalization is today. Social scientists bought in to this paradigm, describing individual investment in education (with professional education at its apex) as "human resources," "human capital," "cultural capital," and movement up the class scale as "social mobility" (Becker 1993 [1964]; Bourdieu et al. 1995 [1964]; Dennison and Gallagher 1986). In practice, however, male-dominated professions paid more on investment than female-dominated ones, and Whites did better than people of colour (Sokoloff 1992). The malestream social science literature was silent on this reproduction of inequity, instead focusing on how neighbouring professions fought over professional turf (Abbott 1988; Gieryn 1983), with an occasional soul-searching about whether professional monopolies really served the public interest (e.g. Freidson 1970; Johnson 1972). Equity issues in professional education were completely ignored until feminists and anti-racists in the 1980s and 1990s started to reveal that professional education was a major mechanism for the reproduction of class, gender (Epstein 1981; Smythe et al. 1999; Witz 1992), ethnoracial, and ableism hierarchies (Pickard 1983; McPherson 1996; Church 1997). Universities as a whole have been critiqued for excluding women and minorities during this period, as well as for fostering the "chilly climate" that awaits any who slip into the academy (Ainley 1990; Backhouse 1988; Dagg 1988; Lorde 1984; Stalker and Prentice 1998).

Appeals to high standards of education and practice have been a perennial aspect of university-based (particularly professional) education, of which the Flexner Report of 1910 was an early example (Axelrod 1990). However, it has been argued that appeals to "excellence" and "quality" similar to those made by transnational corporations have increased in frequency (Readings 1996). Since these discourses of excellence have historically been a justification for exclusion of women and ethnoracial minorities in professional education, it is important to see how they continue to undermine equity initiatives. In my view, it is no accident that many new research programs have the word "excellence" in their title. The use of this word indicates that equity is implicitly being challenged in these discourses. Such globalization discourses also pick up on the "motor of the economy" justification of a generation past; for example, annual reports of the Ontario Centres of Excellence program have suggested that grants for high technology research increased university "productivity," the number of highly-skilled graduates, and the effectiveness of business-university relationships (Bell 1996).

Although some scholars believe that globalization discourse has fundamentally transformed the university from a haven for scholars (e.g. Newson and Buchbinder 1988), historical evidence suggests that academic capitalism always was one of the building blocks of the North American university. Gidney and Millar's account of the process by which the University of Toronto pursued and then incorporated proprietary schools of medicine and law in the late 1800s, for example, shows that Canada's largest university has been a key player in academic capitalism almost from its founding (1994). As early as 1918, Thorstein Veblen in *The Higher*

Learning in America, warned against the incorporation of professional schools into the academy, lest their connections with big business corrupt it.

But is there a new kind of academic capitalism? The retrenchment of state spending on universities and their increased reliance on corporate funding since the 1990s is pinpointed as ushering in new connections between big business and the university. The evidence suggests that at least four processes have changed recently. 1) There has been more direct input into research funding, hiring, and faculty assessment and agendas by transnational corporations (Graham, 1998). Early links have been made between these processes and the reproduction of inequity. For example, the huge gender gap in the naming of research chairs in Canada has been traced to their nomination by university presidents (rather than by peer competition), whose job is corporate fundraising (Begin and Stewart 2002). Also changed is 2) the nature of academic labour itself, which seems to have incorporated a "logic of accounting" into faculty hiring, assessment, and promotion (Broadbent and Laughlin 1997; Currie and Newson 1999; Slaughter and Leslie 1997) and 3) university architecture and capital funding. The newest buildings on Canadian campuses clearly advertise the commercialization of professional schools. In this brave new world, teaching hospitals look like shopping malls and research institutes like mini-bank skyscrapers. Upon entering the Laval campus in Quebec City, one first encounters the Abitibi-Price Pavillion which houses the forestry faculty. At the University of Toronto, the new Rotman building with a façade that juts out towards the main street of the campus contains the business faculty. The new Munk Centre that sits beside Trinity College showcases international relations, the global face of academic capitalism. At the University of Guelph, the home of veterinary science, a new skyscraper-like coloured-glass building contains research facilities for work on genetically modified organisms. But the most important change, and the one that is my main concern in this paper, is 4) the effect of corporate steering on the curriculum.

HOW UNIVERSITY CURRICULA REFLECT CORPORATE INTERESTS AND REPRODUCE INEQUITY

The role of drug companies at universities in Canada has recently been spotlighted by the case of Nancy Olivieri (Thompson et al. 2001). Her writing makes clear the direct links between university research and this industry (Olivieri 2001). My own research over the past ten years has focused on pharmaceutical science education and how it is linked both to pharmacy practice and to the interests of the pharmaceutical industry. My colleagues and I have also explored the "record" in terms of equity, showing mechanisms by which the curriculum subjugates all knowledge except that of interest to drug manufacturers (Muzzin 2001a, b). Briefly, like forestry, pharmacy imparts a curriculum to its students that could have been designed by the transnational (and homegrown) industries whose plaques adorn the walls of its buildings. The pharmacy curriculum focuses almost exclusively on drugs, marginalizing herbal and other originally Indigenous knowledges that were part of its curriculum 100 years ago. An examination of the course calendar shows that students now learn only about pharmacology and molecular biology, which is the basis of the synthetic "designer drugs" that drug companies produce (Muzzin 2001a). As with forestry and plant genetics, where Monsanto designs the crop genetically to make a desired product, the Glaxos and Mercks of the world design drugs that turn on and off our human receptor systems, arguing that herbs are "sledgehammers" while their own designs are fine tools for readjusting the human body. As these corporations redesign life, graduate and undergraduate studies of trangenic processes and products also multiply.

At the undergraduate level, pharmacy is still in the business of training professionals. But pharmacy over the past century has been contested professional territory, with much of its original jurisdiction as the "poor man's doctor" displaced through industry and socialized medicine in Canada (Muzzin et al. 1998). In the interests of promoting the survival of the profession of pharmacy, clinical pharmacy educators have recently introduced what they see as a more client-centred mode of practice which they call "pharmaceutical care" (but which remains biomedically reductionist in its definition of what patients need). Thus although the pharmacy curriculum is corporate-driven, there is bit of a break from a corporate-saturated curriculum and an opportunity, however limited, to talk about "care." The real problem appears only when one examines who teaches what. Of the 320 teachers of pharmacy in Canada in 1996, almost 200 taught clinical topics while the rest taught basic sciences. Only about a quarter of these clinical and social administrative professors were tenured or tenure-stream, while almost all of the basic science professoriate is tenured (Muzzin 2001b). This means that the "caring" or professional practice component of the curriculum is not as firmly ensconsed in the academy as the scientific research aspect. Predictably, almost 90 percent of the basic science professors are men, while three-quarters in the clinical stream teach part-time or contractually, the vast majority of whom are women. Thus women – the familiar "reserve army" of labour – are not very involved in the basic research agenda that dominates the curriculum and they are easily replaced in their secondary function of training professionals. This curriculum thus reproduces gender inequality, subjugates discourses of practice, and serves to disguise the real university business of corporate-friendly research, much as Bloom (1988) argues that the practice-based curriculum in medical education is just "window dressing" for molecular research.

The generalizability of this specific case of academic capitalism to other sciences and professions has emerged in ongoing research on equity issues across a variety of university-based professional fields including dentistry, optometry, social work, education, law, and nursing.[2] Much of the curriculum of these professions is either technologically reductionist or a confusing mix of social control and emancipatory discourses, as critics of professional education have pointed out. However, behind the façade of these contested curricula lies the mechanism that lodges these professional faculties so firmly in the restructured academy – research funding that explores topics of interest to transnationals. This agenda sometimes involves basic researchers in, say, dentistry or nursing teaming up with biomedical researchers, even when this research seems irrelevant to the particular faculty within which it is housed. Second, a few basic scientists in fields such as dentistry told me that they had given up trying to access government funding councils (where they don't fare as well as researchers in medicine) and were taking pharmaceutical firm money directly. Similar arrangements exist between contact lens manufacturers and laser surgery firms and optometry faculty. The bottom line is that when the researchers bring in research money, the university administration is impressed, and some of the pressure to cut back on very expensive professional education is eased.

Although I have emphasized the gendered nature of this process, it is important to see that it is also racialized. This is more difficult to detect, though the fact that there are so few people of colour teaching in the academy is a big clue. In science and professional faculties, an occasional man of colour may find his way into the academy via the prestigious route of biomedical research. However, this career path typically involves indenture to a senior White male academic scientist for many years without promise of a future position. Through strategic alliances with senior

White male scientists, he may be able to achieve tenure. I have encountered few such individuals over the years, none arriving without significant struggle. At the intersection between race and gender, there are vanishingly few individuals in the academy. As has been shown many times, women are doubly disadvantaged because they face partriarchal relations both within the academy and within their own families and communities. I once interviewed a couple of colour – a very rare situation – and observed how the man later became a full professor by aligning himself with an old, White funding "star," while his wife spent her time looking after the children, sacrificing her own career in science, and eventually becoming a part-time faculty member. When tenured basic science and law professors obtain grants, often this research is both relevant to promotion of a biomedical or corporate law curriculum and agenda as well as impressive to administrators who do comparative calculations before making restructuring decisions. This process favours basic science or corporate law professors who tend to be White men as well as furthering the neocolonizing agenda that Shiva and other advocates for Indigenous knowledges have critiqued. By my reckoning, the process is likely to accelerate as restructuring in higher education proceeds. Specifically, even though a few White women might have been hired a decade ago to teach clinical topics or even do basic science research, in the face of restructuring, faculties all cite the need to bring in a couple of young, typically White male, "research stars" to "hold the fort." Thus the dismal record of equity in professional education continues and accelerates under retrenchment.

THE UNIVERSITY AS A SITE OF ACTIVISM

There has been a great deal written in the professions literature recently about the proletarianization and deskilling of professionals in advanced capitalism, including the decreasing influence of the professoriate (e.g. Coburn 1983; Coburn et al. 1999; Newson 1992; Rhoades 1999). While it may be true that professionals including academics are losing some of the power they once had, I think that more attention needs to be focused on two aspects of the process. First, following my argument above that academic processes are gendered and racialized, it needs to be emphasized that some professors (White male research stars supporting globalization initiatives?) have as much or more power than they ever did, while the 'masses' of their colleagues doing 'scut work' do not. That is, the university's dismal record on equity is part and parcel of its involvement in the promotion of globalization, and the so-called proletarianization is not uniform.

Second, I think that professions research should focus more directly on the *university* and its knowledge-production machine as opposed to the workplace as the main site at which professional inequities are produced. This is not to say that inequities in the workplace are unimportant. However, as professors, we have direct access to university classrooms, admissions and search committees, and conferences where the centre of the marginalization that I have described above is taking place, and the opportunity to document it. Also, to promote equity and environmentalism, we have opportunities to insert courses demonstrating the reproduction of gender, ethnoracial, and other difference as well as the downgrading of Indigenous knowledges along with the environment, within the very programs that we are teaching. Those of us with tenure and academic freedom also can expose the excesses of academic capitalism, as OISE/UT has allowed a colleague and me to do in a course with this name. We can hold out for the hiring of women and ethnoracial minorities on search committees, and publish the details of our insights about promoting equity and environmental activism. It must be acknowledged that activism faces backlash both in the academy and the media (see Blackmore 1997) but we are in the best position to put forward this important agenda.

It may be difficult, but is not impossible, for the tenuously employed part-time or contractual faculty who have been described here to address equity issues in their teaching and publications. One of the difficulties arises from the consistently higher teaching loads for full-time clinical or "teaching stream" faculty as compared to basic science faculty, as my research shows. Simply put, higher teaching loads keep the few full-time professional practice professors busy and free the scientists to devote their time to furthering the cause of biotechnological or other capital-friendly research. Second, particularly within reductionistic professional schools which do not support critical scholarship, it may be difficult to raise issues of equity without being marginalized further as promoting self-interest. But with the assistance of networks of critical scholars, such as those who generously supported me when I was in pharmacy, even this may be possible. The existence of courses on professional ethics and practice provide many such opportunities.

Contingent faculty are the largest growing sector of the university, and the mechanisms of inequity driving this system of retrenchment as part of globalization need to be problematized by tenured faculty writing about the university. As well, university faculty associations and unions need to be made aware of, and work on, these problems.

Perhaps the most difficult nexus of all is the intersection between exclusion and marginalization in hiring that goes hand in hand with the subjugation of the activities and epistemologies associated with women and ethnoracial minorities. White feminists have been particularly sensitive to the marginalization of "caring" discourses in the academy. And there is evidence that historically, women physicians questioned the medical discourses in the nineteenth century about women's "insanity" and "nervousness" and offered alternative ways of constructing women and gender (Theriot 1996). What is now subjugated knowledge across all the professional fields is the labour-intensive teaching of practice or community knowledge that was so dominant at the turn of the previous century and that predated the incorporation of professional practice into the university. These are the very practices that Schön (1983) and others have suggested need to be "shored up" in the academy, if only to address the crisis of the mindlessly technical nature of professional education. Some of these practices, such as herbal knowledge, were originally Indigenous, as argued by Ainley in this volume. I think that these knowledges and practices are daily brought right into the academy via part-time faculty who regularly engage in community teaching, healing, and cultural activities (e.g. native healing circles), but as I have emphasized, their positions are not secure.

Broadbent and Laughlin (1997) have argued that alternatives that emphasize the needs of the community can be maintained by what they label an "absorbing group." I observed this among nursing professors who teach therapeutic touch on a volunteer basis in their communities, although they were not being appropriately remunerated for this work. Without academic recognition, absorbing groups are hard to maintain. Segregated into part-time work, the equivalent of seasonal farm workers are not true "academic citizens" in the disciplines they inhabit and thus typically have limited power to add important topics permanently to the curriculum. However, students do benefit from their labours. When seen from this perspective, the proletarianization of professionals can be seen to have obvious ethnoracial aspects as well, since these local knowledges are continuously replaced with the knowledges of White technical science and technical law. However, bringing these knowledges into the academy is a significant source of resistance and deserves the support of full-time faculty.

CONCLUSIONS

I have argued here that the brave new world of professional education in Canada is reflected in the coloured glass of its corporate façades. These façades, built by universities with industry and government money, advertise the knowledge produced within them. An examination of the curricula and policies of these professional schools shows that they promote the production of trees for market, consumption, and profit rather than teaching about the spirituality of the forest (*Asking Different Questions* 1996); enshrine intellectual property rather than sharing of research results (University of Toronto 2000); endorse globalization and technology transfer of White engineering to the rest of the world rather than protecting Indigenous knowledges (Shiva 1997); and encourage the control and manipulation of natural organisms and ecosystems rather than the appreciation of their beauty and complexity (Tokar 2001). In summary, universities are institutions that reproduce hegemonic transnational interests. One way this occurs is through a science enterprise driven by academic capitalism that hides behind the training of university-level professionals. Although often irrelevant or contradictory to professional practice, this capital-friendly mode of knowledge production insinuates itself and saturates professional curricula, marginalizing the practice-based knowledges that preceded technoscience and technolaw. Non-tenured faculty who teach professional practice are marginalized by their tenuous positions at the academy, while the science that drives globalization is ensconced through the tenure system. Finally, this system is gendered and racialized through the androcentric Whiteness of university research stars and the transnational-friendly research that they promote in the restructured academy.

I end upon a note of hope – the classroom, committees, conferences, and publications have become a site of activism for me, as they have always been for responsible academics. But don't take this as a sugar-coating. I would definitely prefer it if the university I know didn't exist and if the history of the university could have delivered the emancipatory promise that it pretended to – for real and not only between the lines of a hegemonic text.

NOTES

1 This is a line from Tracy Chapman's song, "Tell it like it is."
2 Over 200 full-time faculty in education, pharmacy, dentistry, and social work

were interviewed between 1995 and 1998 as part of SSHRC Grant, "Making a Difference" to Sandra Acker (P.I.), Linda Muzzin, Carol Baines, Marcia Boyd, and Grace Feuerverger. About 160 non-tenure stream faculty have been interviewed since 1999 in pharmacy, dentistry, law, optometry, and nursing as part of a SSHRC grant, "Gendered Retrenchment," to Linda Muzzin (P.I.), Marcia Boyd, Fran Gregor, and Marlee Spafford, with collaborators, Cheryl Albas, Avis Mysyk, Vicki Nygaard, and Maria Wallis.

REFERENCES

Abbott, A. 1988. *The system of professions*. Chicago: University of Chicago Press.

Ainley, M. 1990. *Despite the odds. Essays on Canadian women and science*. Montreal: Vehicule.

Asking different questions. Women and science. Toronto: Artemis Films/National Film Board. G. Basen and E. Buffie (directors). Speaker: Peggy Tripp.

Axelrod, P. 1990. *Making a middle class: Student life in English Canada during the thirties*. Montreal/Kingston: McGill-Queens University Press.

Backhouse, C. 1988. *Women faculty at University of Western Ontario: Reflections on the employment equity award*. University of Western Ontario: Faculty of Law.

Barlow, M. and E. May. 2000. *Frederick Street: Life and death on Canada's Love Canal*. Toronto: HarperCollins.

Becker, G. 1993 [1964]. *Human capital. A theoretical and empirical analysis with special reference to education*. Third Edition. Chicago: University of Chicago Press.

Begin M and M. Stewart. 2002. *Women and health*. Presented at the 12th international conference of women engineers and scientists (ICWES12), Ottawa Congress Centre, Ottawa, Canada, July 27–31.

Bell, S. 1996. University-industry interaction in the Ontario centres of excellence. *Journal of Education* 67: 322–48.

Bertell, R. 2001. *Planet earth: Latest weapon of war*. Montreal: Black Rose Books.

Bevis, O.O. and J. Watson. 1981. *Toward a caring curriculum: A new pedagogy for nursing*. New York: National League for Nursing.

Birke, L. and R. Hubbard. 1995. *Reinventing biology. Respect for life and the creation of knowledge*. Bloomington: Indiana University Press.

Blackmore, J. 1997. Disciplining feminism: A look at gender-equity struggles. In *Dangerous territories: Struggles for difference and equality in education*, eds. L.G. Roman and L. Eyre. New York and London: Routledge.

Bloom, S. 1988. Structure and ideology in medical education: an analysis of resistance to change. *Journal of Health and Social Behaviour* 29 (Dec.), 294–306.

Bourdieu, P. and J. Passeron. 1990. Reproduction in education, society and culture (2nd ed., R. Nice, Trans.) London: Sage.

Bourdieu, P., J.-C. Passeron and M. de Saint Martin. 1995 [1964]. *Academic discourse*. Stanford, CA: Stanford University Press.

Broadbent, J. and R. Laughlin. 1997. Accounting logic and controlling professionals. In *The end of the professions? The restructuring of professional work*, eds. J. Broadbent, M. Dietrich and J. Roberts, 34–9. London: Routledge.

Brown, E.R. 1979. *Rockefeller medicine men*. Berkeley: University of California Press.

Burt, S. and L. Code. 1995. *Changing methods. Feminists transforming practice*. Peterborough: Broadview Press.

Church, K. 1997. *Forbidden narratives*. Gordon and Breach Publishers.

Clark, E.A. 2001. GM foods. Why and why not? Presented at the *Teaching As If The World Mattered* Conference, OISE/UT, May 11–15.

Coburn, D. 1983. "Medical dominance in Canada in historical perspective: the rise and fall of medicine?" *International Journal of Health Services* 13(3), 407–32.

Coburn, D., S. Rappolt, I. Bourgeault and J. Angus. 1999. *Medicine, nursing and the state*. Toronto: Garamond Press Ltd.

Currie, J. and J. Newson, eds. *Universities and Globalization*. Thousand Oaks, CA: Sage.

Dagg, A.I. 1988. *MisEducation: Women and Canadian universities*. Toronto: Ontario Council on Graduate Studies.

Dennison, J. and P. Gallagher. 1986. *Canada's community colleges: A critical analysis*. Vancouver: University of British Columbia Press.

Epstein, C. 1981. *Women in law*. Urbana, IL: University of Illinois Press.

Franklin, U. 1999 [1990]. *The real world of technology*. Revised Edition, Toronto: Anansi.

Freidson, E. 1970. *Professional dominance*. Chicago: Aldine.

Gaffield, C. and Pam G. 1995. *Consuming Canada. Readings in environmental history*. Mississauga: Copp Clark Ltd.

Gieryn, T.F. 1983. Boundary-work and the demarcation of science from non-science: Strains and interests in professional ideologies of scientists. *American Sociological Review* 48: 781–95.

Gidney, R. and W. Millar. 1994. *Professional gentlemen. The professions in nineteenth century Ontario*. Toronto: University of Toronto Press.

Good, B. and M.-J.D. Good. 1993. Learning medicine: The construction of medical knowledge at Harvard medical school. In *Knowledge, power and practice*, eds. S. Lindebaum and M. Lock, 81–107. Berkeley: University of California Press.

Graham, W. 1998. "Corporatism and the university. Part 2." *University of Toronto Faculty Association Newsletter*, 28 Feb.

Haraway, D. 1978. Animal sociology and a natural economy of the body politic, part II: The past is the contested zone. *Signs*. 4(1), 37–60.

- 1991. *Simians, cyborgs and women*. New York: Routledge.
- 1997. *Modest witness@second millennium: FemaleMan© meets oncomouse.™ Feminism and technoscience*. New York: Routledge.

Harding, S. 1986. *The science question in feminism*. Ithaca: Cornell University Press.
- 1998. *Is science multicultural? Postcolonialisms, feminisms and epistemologies*. Bloomington: Indiana University Press.

Hubbard, R. 1995. *Profitable promises. Essays on women, science and health*. Monroe, MN: Common Courage Press.

Johnson, T. 1972. *Professions and power*. London: Macmillan Press.

Keller, E.F. 1996. Language and ideology in evolutionary theory: reading cultural norms into natural law. In *Feminism and science*, eds. E. Keller and H. Longino, 154–72. Oxford: Oxford University Press.

Lorde, A. 1984. *Sister outsider*. The Crossing Press.

Marchak, M.P. 1983. *Green gold: The forest industry in Canada*. Vancouver: UBC Press.
- 1995. *Logging the globe*. Montreal and Kingston: McGill Queen's University Press.

Martin, E. 1996 [1991]. The egg and the sperm: How science has constructed a romance based on stereotypical male-female roles. Reprinted in *Feminism and science*, eds. E.F. Keller and H. Longino, 103–17. Oxford: Oxford University Press.

McNeill, J.R. 2001. *An environmental history of the twentieth-century world*. New York: W.W. Norton and Co.

McPherson, K. 1996. *Bedside matters*. Toronto: Oxford University Press.

Middleton, C. 2000. *Education for sustainability in the built environment: An inquiry into change in architecturally related curricula in Ontario universities*. EdD Thesis Draft.

Muzzin, L., G. Brown and R. Hornosty. 1998. Professional ideology in Canadian pharmacy. In *Health and Canadian society*, eds. D. Coburn, C. D'Arcy, and G. Torrance, 379–98. Toronto: University of Toronto Press.
- 2001a. Academic capitalism and the hidden curriculum in the pharmaceutical sciences. In *Unhealthy times*, eds. P. Armstrong, H. Armstrong, and D. Coburn, 97–120. Don Mills, ON: Oxford University Press.
- 2001b. 'Powder puff brigades': Professional caring vs. industrial research in the pharmaceutical sciences curriculum. In *The hidden curriculum in higher education*, ed. E. Margolis, 135–54. New York: Routledge.

Newson, J. 1992. The decline of faculty influence: Confronting the effects of the corporate agenda. In *Fragile truths. Twenty-five years of sociology and anthropology in Canada*, eds. W. Carroll, L. Christianson-Ruffman and D. Harrison, 227–46. Ottawa: Carleton University Press.

Newson, J. and H. Buchbinder. 1988. *The university means business.* Toronto: Garamond.

Noble, D. 1977. *America by design. Science, technology and the rise of corporate capitalism.* Oxford: Oxford University Press.

Noske, B. 1997. *Beyond boundaries. Humans and animals.* Montreal: Black Rose Books.

Olivieri, N. 2001. When money and truth collide. In *The corporate campus*, ed. J. Turk, 53–62. Toronto: James Lorimer and Co.

Pickard, T. 1983. Experience as teacher: Discovering the politics of law teaching. *Univ. Toronto Law J.*, 279–314.

Readings, B. 1996. *The university in ruins.* Cambridge, MA: Harvard University Press.

Rhoades, G. 1999. *Managed professionals. Unionized faculty and restructuring academic labour.* New York: SUNY Press.

Schön, D. 1983. *The reflective practitioner.* New York: Basic Books.

– 1987. *Educating the reflective practitioner.* San Francisco: Jossey-Bass.

Shiva, V. 1988. *Women, ecology and development.* London and Dehli: Zed Books and Kali for Women.

– 1995. *Monocultures of the mind. Perspectives on biodiversity and biotechnology.* London: Zed Books.

– 1997. *Biopiracy. The plunder of nature and knowledge.* Boston: South End Press.

Slaughter, S. and L. Leslie. 1997. *Academic capitalism. Politics, policies and the entrepreneurial university.* Baltimore: Johns Hopkins.

Smythe, E., S. Acker, P. Bourne and A. Prentice (eds.) 1999. *Challenging professions.* Toronto: University of Toronto Press.

Sokoloff, N. 1992. *Black women and white women in the professions. Occupational segregation by race and gender, 1960–1980.* New York: Routledge.

Stalker, J. and S. Prentice. 1998. *The Illusion of inclusion. Women in post-secondary education.* Halifax: Fernwood Press.

Theriot, N. 1996. Women's voices in nineteenth-century discourse: A step toward deconstructing science. In *Gender and scientific authority*, eds. B. Laslett et al., 124–54. Chicago: University of Chicago Press.

Thompson, J., P. Baird and J. Downie. 2001. *The Olivieri report. The complete text of the report of the independent inquiry commission by the Canadian Association of University Teachers.* Toronto: James Lorimer.

Thompson, S. 2002. *Environmental justice in a toxic economy: Community struggles with environmental health disorders in Nova Scotia.* PhD Thesis, OISE/University of Toronto, Department of Adult Education, Community Development and Counselling Psychology.

Tudiver, N. 1999. *University for sale: Resisting corporate control over Canadian higher education.* Toronto: James Lorimer.

Tokar, B. 2001. Ed. *Designing life?* Montreal and Kingston: McGill-Queen's University Press.

Turk, J. 2000. Ed. *The corporate campus.* Toronto: James Lorimer.

University of Toronto. 2003. Policies: Research – intellectual property. http://www.library.utoronto.ca/techtran/summary.html. (Retrieved November 13, 2003.)

Veblen, T. 1968 [1918]. Fifth printing. *The higher learning in America.* New York: Hill and Wang American Century Series.

Witz, A. 1992. *Professions and patriarchy.* London: Routledge.

11

Working in the Field of Biotechnology

ELISABETH ABERGEL

The downstream effects of scientific research, as Martha Crouch once observed, are seldom assessed and reflected upon by scientists, for the idea of the neutrality of scientific results persists (Crouch 1991). While sociologists, feminist scholars, and philosophers of science have convincingly uncovered the contingent and constructed nature of scientific knowledge and its production, most practising scientists are unlikely to have encountered these important works and reflected upon their implications for the conduct of their discipline. Context specificity is a crucial aspect of all knowledge production and in particular for framing knowledge claims; its inclusion in scientific experiments is believed to introduce 'bias' in experimental research and therefore, any discussion of context constitutes 'flawed science.'

The independence of social, gender, class, race, historical, cultural, environmental, and other contexts from the conduct of experimental science, or the 'detachment' of subject from object, is commonly used to defend the objectivity and neutrality of science and scientific facts. Scientific knowledge should theoretically be valid independently of one's gender, race, social, or geographical location. Nevertheless, scientific knowledge – the great equalizer, as Enlightenment thinkers believed – has not yet successfully broken down the structural and ideological barriers faced by women entering scientific professions. Unfortunately, women are vastly absent from certain scientific disciplines and are disproportionately underrepresented in the upper reaches of their profession in fields where they have successfully participated (Etzkowitz et al. 2000).

Moreover, scientific concerns and research directions do not reflect issues that affect women's lives.

OBJECTIVITY AND SUSTAINABLE SCIENCE

I argue here that context specificity is a vital aspect of any feminist scientific worldview, and that whether it is described as "standpoint epistemology" (Harding 1996) or "situated knowledge" (Haraway 1991), such reflexive practice on the part of the scientific community would go a long way towards contributing to a more sustainable scientific system (Schiebinger 1997). It should in fact be included in any discussion of research methodology and in any socially responsible or sustainable scientific enterprise. It would also contribute to a more democratic science where research and knowledge claims could be arrived at through a shared process involving communities.

Historically, the objectivity of science has been naturalized and normalized into scientific and research practices, so the primary task of feminist scientists was to de-naturalize and de-normalize such claims. While the first feminist studies of science focused on the underrepresentation of women in the various fields of scientific inquiry, this was followed by the historical 'recuperation' of women scientists who had been intentionally left out of the historical record. In general, feminist scholars argue that Western science is the product of a conceptual worldview that incorporates at its roots, specific ideologies of gender (Keller and Longino 1996). The study of these entrenched epistemological and social practices allowed for the conceptual deconstruction of scientific theories and institutions, bringing to light the power structures contributing to women's oppression and the devaluation of their knowledge. Genderized biases reinforced the idea that science was a powerful way to rationalize inequity and specifically exclude women and whole segments of humanity from full participation in society. While these early studies shed light on many important questions for feminists, they failed to confront the real questions in terms of how the practices of Western science were directly affecting women's lives. Thus, scholarship in feminist science studies has progressed from the analysis of science *about* women, to women *in* science and, finally to science *for* women, where current works centre around redefining women's relationship to science and knowledge systems that integrate women's everyday realities and experiences. Schiebinger's call for a more sustainable science resonates, as feminism is no longer simply a tool for critique but rather, a rich resource for generating new knowledge (1997, 3). In this sense, the current feminist program in science studies differs from traditional science studies where feminism seeks to propose new visions for science and for women scientists. The research network, "Biology As If The World Mattered" (BAITWORM), which the editors of this book founded, has created a feminist space for discussing scientific practice and for exploring the critical elements left out of scientific projects and out of women's lived experiences of science. Involvement in this group transformed participants from passive roles as scientists or science scholars to more active ones as advocates for social change. This paper aims to retrace some of my own thinking and personal experiences, having had the privilege to explore some of these issues as a member of BAITWORM.

I was trained as a molecular biologist and practised biopharmaceutical research in an industrial setting. After some mixed experiences in scientific research, I pursued a doctorate in Environmental Studies, reflecting on my work as a scientist and on what my scientific training and genetic experiments had taught me about the nature of science and the science of Nature. Much of my academic and activist critique derives from my experience in the biotechnology industry, having had the opportunity to witness first-hand how scientific utopia converged with neoliberal ideology. In this brief article, I will offer some critical thoughts about following the products of scientific research in plant biotechnology out into the world to see how they are used and how they contribute to social, cultural, and environmental change. I use some of my work and thoughts on Genetically Modified Organisms (GMOs) to reflect on some of these ideas.

SCIENCE REVISITED

The transition from a scientific understanding of the environment to an environmental understanding of science entailed a rejection of some of the principles that guided my work as a bench scientist; my values and belief system. During my scientific experiments, I was searching for ways to make my manipulations of living organisms socially and environmentally relevant. Most of my scientific training painted a picture of science and scientists (mostly European males) as 'pioneers' uncovering natural phenomena through a linear and cumulative process of discovery. This approach of thinking of the world as knowledge waiting to be uncovered and manipulated through human ingenuity is a dominant feature of biotechnology (or "secrets of life") research.

However, it is now widely accepted that the legitimacy of scientific explanations does not simply originate from their ability to provide the best account of reality but largely from their suppression of alternative systems of knowledge. Since the Enlightenment, science has been a dominant and dominating discourse, serving the interests of powerful elites. Science expressed *through* the application of the scientific method is often presented as objective, truth-seeking, and value-free. Yet, knowledge is not neutral; it is context- and framework-dependent. Sandra Harding (1998) has argued that conventional accounts of Western science have tended to unify knowledge about nature as 'one nature,' 'one truth,' and therefore 'one science.' This perspective has contributed to an understanding of the sciences within a larger global context, charting a map for European sciences within a global economic and political order and relating it (in destructive and exploitative ways) to non-European cultures and sciences.

It follows that in order to construct a more 'sustainable' science, scientists and society at large must examine how science is organized and scientific claims are used to produce more wealth for some and total deprivation for others. It becomes essential to grasp how chance and diversity in Nature become part of an ordered totality, one that is subjective, value-laden and Eurocentric. On the other hand, for Harding and other feminists, it is important to situate the problematic of scientific objectivity, not in totalizing discourses or grand narratives of science, but in the everyday lives of women who have been marginalized. It is the view from below that informs and guides feminist research.

For many who critique science, the scientific enterprise embodies the triumph of the masculine over the feminine, of mind over matter, of the "West over the Rest," of human ingenuity, and Capital over Nature – all of which are clearly expressed in the biotechnology project. Above all, science is incapable of constructive self-reflexivity; it resists exposing itself up to the kind of critical examination and ethical scrutiny it demands of its subject matter. On this point it is limited and limiting as a worldview and as a way of dealing with social, ethical, environmental, and political issues.

WINNERS AND LOSERS

Controversies surrounding the large-scale environmental releases of GMOs and their subsequent integration into our food system highlight the risks and uncertainties posed by this new technology and the socio-ethical questions arising from technoscience. The issues emerging from agricultural biotechnology are multidimensional, but have been largely framed as technical or scientific in nature. In fact, opposition to GMOs has often been portrayed as characterizing the public misunderstanding of science. The lack of scientific literacy on the part of the general population is considered a systemic problem at the root of an increasing anti-science sentiment. The public misunderstanding of science argument works both ways: on the one hand to confound an unsuspecting public and promote the biotechnology project as a "public good" and, on the other hand to silence its critics. The argument that if people could understand the science, they would embrace the technology, ignores more profound questions surrounding the industrial exploitation of living organisms and the directions in which technoscience is evolving. The blurring of boundaries between micro-organisms, plants, animals, and humans is just one of the many aspects of biotechnology that require careful consideration. By asking questions about who the science is for; where it comes from and why it is necessary; whether low-cost, sustainable alternatives exist; and who benefits and who doesn't, scientists can engage in reflexive practice and incorporate social responsibility into their work. Another important point is the proximity of researchers to the recipients (the knowledge users) of the research results; the closer the involvement of researchers in the community, the more likely scientists can witness the consequences of their work and become part of a research network that includes members of the community as potential co-researchers. These interactions make for a more socially responsible science. Because genetically-engineered organisms in the form of crop plants originate in the laboratory and not the farm, they should be considered suspect.

SOME OBSERVATIONS ON ASSESSING THE ENVIRONMENTAL SAFETY OF GMOS

One of the ways GMOs have been able to be commercialized is through establishing their safety in the environment, once they are released. This has been facilitated through the development of government regulations based on the scientific evaluation of potential ecological risks. The field of risk assessment for testing the environmental safety of GMOs is fraught with scientific controversy. At its root, it pits a reductionistic account of biological processes, mostly advocated by molecular biologists, against a more holistic approach defended by ecologists and evolutionary biologists. Since ecological thinking has played a marginal role in the biosafety debate, and ecological literacy remains relatively low, the assessment of risk has been narrowly defined to accommodate the rapid commercialization of genetically engineered (GE) crops. It follows that scientific controversies over genetically engineered organisms tend to reflect disputes over worldviews and divergent conceptions of science rather than technical issues *per se* (Regal 1998).

The fact is that the commercial time frame used to evaluate the risks posed by GE crops cannot uncover long-term ecological or human health impacts. In addition, the determination of risk is left in the hands of scientific 'experts' and the developers of new crops. In theory, environmental risks are distributed among people and therefore everyone has a stake in defining and managing them (Beck 1996). The scientific determination of ecological risks has failed to account for different conceptions and value-frames. Lay knowledge about the environment and health has been shown to be a precious asset in scientific research.

Public involvement in decision-making should be an integral component of any policy regarding the public funding of genetic engineering research. Unfortunately, the precautionary approach advocated by many is incompatible with the race for profits, which demands the rapid release of GE products into the marketplace. Governments, through financial incentives and flexible regulatory regimes, have *de facto* enabled the proliferation of biotechnology in agriculture. Unsuspecting consumers might be tempted to reflect on the way in which GE products are being enforced – no clear labels on products containing (and not containing) GE ingredients despite repeated public demands; no visible campaigns or efforts to inform the public about what they are purchasing; no public dialogue or debate on people's lack of desire to consume GE foods; a strategy of placing biotechnology on a natural continuum with traditional food production methods (such as cheese making or beer brewing), in order to defuse any claims that GE foods might not be safe or wholesome; the suppression of alternative scientific viewpoints; the consistent downplaying of scientific uncertainty in government decision-making. What is obvious is that the lack of public awareness actively sought by proponents is equated with the tacit acceptance of GE foods by people. What survey after survey has proven to be counterintuitive to the majority of people has been deemed desirable by government s committed to industrial growth.

THE NATURE OF AGRICULTURAL BIOTECHNOLOGY

Biotechnology functions by defining 'needs' that are rooted in an industrial vision of Nature and biological processes. Biotechnology inhabits a moral and physical landscape where depleted resources, barren lands, and polluted air and water directly correlate with overpopulation, disease, and starvation as the inevitable by-products of human progress. The starting point for biotechnology is the tacit acceptance of environmental degradation and resulting social crisis. Biotechnology becomes the coping mechanism that claims it is "business as usual" and precludes altering our way of life as well as searching for possible sustainable alternatives. This apocalyptic vision of the world makes it seem like the genetic manipulation of living organisms is the only way out of our evolutionary impasse. It provides the rationale for the manipulation of living organisms and the continued exploitation of natural resources. As humans have learned to overcome the vagaries of Nature, they have placed demands on environments that have rendered them 'deficient' and limited for industrial society's purposes and insatiable appetite for more resources. The efforts put into creating 'salt-tolerant' or 'drought-resistant' plants are justified by defining salinity and drought as genetic problems. Deficiency discourse works by reframing living processes as inherently "improvable," hence perfectible through science at the molecular level. For the genetics industry, genes are where problems arise and where they can be solved. After all, it is easier to disrupt plant and animal genomes than it is to dismantle the political, industrial, and economic order. Clearly, the solution becomes the problem.

BREEDING RESISTANCE

Our reliance on expert knowledge, elite crops, and the corporate control of our genes place us in a relationship of dependency. Power gets distributed from the top down and people have few opportunities to legitimately participate in important societal choices. Biotechnology is restricted to the field of experts where citizens are effectively excluded from meaningful engagement and relegated to passive consumerism. Resistance to biotechnology gets defined narrowly as a contest between Ludditism and techno-scientism. However, things are never as simple as they appear. Pro-biotechnology forces are quick to blame the lack of public confidence in GE on 'public ignorance' and slow to acknowledge that their 'aggressive marketing techniques' contribute to heightened public unrest. In addition, there is a fine line between 'educating the public on the benefits of GE foods' and force-feeding people false hopes and promises.

State support for biotechnology has truly confounded public and private interests and put into question the government's ability to protect the well-being of its citizens. This conflation of interests explains why public concerns have been addressed superficially long after GE products were introduced on the market. Once biotechnology becomes an economic reality, it is increasingly difficult to contest its legitimacy and to find ways to stop it. Ironically, the reluctance of governments and industry to acknowledge the inherent risks of biotechnology has largely contributed to the public unrest and distrust.

It is becoming clear that while biotechnologists were breeding resistance into plants, they also inadvertently bred a different kind of resistance. Europeans have actively resisted GE food and their actions have galvanized public attention worldwide. Public support for organic production is gaining ground as people realize that investing in environmentally friendly agriculture extends beyond simple commodity relations. The anti-GE food campaigns are increasing everywhere as global awareness is raised. Reconnecting people and their food is an important link in the efforts to counter the biotechnology onslaught. Public awareness is the first step towards political action and towards reclaiming a public role in determining the direction of science.

The degradation of environmental systems and the failure of democratic institutions provide the rationale for new technologies such as biotechnology to emerge as the vital links to a 'sustainable' future. We are told the environmental crisis can be overcome by 'improving nature' and democracy can be ameliorated through more choices in the marketplace or through information technologies (genetic engineering is often conflated with information technologies: DNA is described as "software" and cells as "hardware"). This reveals a particular vision of human progress which leaves out most of humanity, ignores the fact that ecocidal practices are still encouraged, and pretends that corporate values are synonymous with democratic principles. In fact, in a bizarre twist, the same multinational agro-chemical corporations directly implicated as the main culprits in the degradation of environmental and human health are now poised to save the planet through genetic engineering. Through the co-optation of environmental and other marginalized discourses, the multinational corporations involved in producing toxic chemicals and imposing unsustainable farming practices are now calling themselves 'environmental companies.'

This gives rise to what has been called an "epochal shift" by which environmental and social problems get turned into opportunities for these industrial interests to invest in more scientific research and development and to achieve greater technological control (Levidow 1999). Many critics of biotechnology have suggested that the genetic manipulation of organisms does little to affect problems at their root causes; instead, attention should be focused on discouraging systemic practices that render the planet and its people sick. The reduction of Nature and organisms to simple "genetic production units" goes against the search for holistic, diverse, equitable, and viable options for saving the environment and improving people's lives. Biotechnology creates organisms and new sets of conditions with no ultimate ecological purpose other than feeding industrial capital. Agricultural technologies come out of the laboratory, not the farm. Decisions about the direction of agriculture are made in the boardrooms of transnational corporations and are facilitated by consenting governments, not through farmer or citizen input. Farmers, like scientists, once independent thinkers, are now seeing their autonomy restricted by industrial pressures to be productive and competitive. Funding research into high-input, monocultural, and highly technological agricultural systems excludes competing notions of agriculture and science so that agricultural biotechnology emerges as the only viable model of food production. Monoculture threatens scientific diversity. Genetic engineering targets environmental and social problems at the genetic level eluding larger ethical questions. Efficiency and competition are the organizing principles in the development of genetic engineering as a "social project." Through this process, market logic overrides ethical considerations and democratic values.

GETTING THE MOST OUT OF LIFE

Biotechnology commits us to a single vision of society where corporate values dominate, threatening notions of social progress and environmental justice. This vision undermines democratic processes and reduces citizens to mere consumers whose rights are determined by their purchasing power. Our relationship to the land, in the case of agriculture, becomes exclusively economic. Bourdieu observes that neoliberalism works by orienting the economic choices to those who dominate economic relationships. In this sense, genetic technologies conform to

Bourdieu's assessment, since those who already dominate economic relationships are allowing the intrusion of multinational corporations into every aspect of life: food, health, human reproduction, and justice, to name a few. Genes are the bottom line and their 'value' is determined by their functionality within the logic of industrial production. Genes and their expression products (proteins) are decontextualized from their original source organism and dehistoricized from their ecological and evolutionary role through laboratory manipulations. Testard and Godin have described the practitioners and promoters of the industrialization and exploitation of genetic processes as "molecular conquistadors" (2000).

The understanding of organisms at the molecular level reduces them to a series of biochemical processes, removing them from the complex web of interactions within which they evolved and thrived. The more that is uncovered about an organism's intimate workings at the molecular level, the less relevant its social, ecological, and cultural roles become. It becomes merely a production system or a tool. It is reduced to a "technology." An important lesson I learned throughout my laboratory career was that by taking the genes out of life, we were also taking the life out of genes.

The relentless pursuit of profit through the control of Nature and of the last remaining enclaves of wilderness and biodiversity comes at the expense of human rights and people's self-reliance. Dependency is a by-product of the unhinging of life processes from their contextual and historical past, and the displacement of plants, animals, and rural people is a key element in this process. Neoliberalism also promotes the destruction of collective structures through the devaluation of the public realm, according to Bourdieu, and through what he calls 'moral Darwinism' where the struggle of all against all prevails (1998). As an industry and a set of technologies, agricultural biotechnology exemplifies the struggle for the privatization of public goods such as seeds, genetic resources, and the knowledge that they contain. Through a complex system of intellectual property rights, biotechnology recasts public resources as 'value-added' private goods; Vandana Shiva (1993) claims that the appropriation of genetic resources by multinational agro-chemical companies turns renewable resources into non-renewable ones, where increasingly technologies are devised to replace and supplant people's free access to resources, such as plants for food and medicine. This is done at the expense of local economies, food security, traditional cultural practices, community self-determination, and sustainable scientific practice.

REFERENCES

Beck, U. 1996. *Ecological enlightenment: Essays on the politics of the risk society.* Atlantic Highlands: Humanities Press International, Inc.

Bourdieu, P. 1998. L'essence du néolibéralisme. *Le monde diplomatique.* March.

Crouch, M. 1991. The very structure of scientific research mitigates against developing products to help the environment, the poor, and the hungry. *Journal of Agricultural and Environmental Ethics,* 151–65.

Etzkowitz, H., C. Kemelgor and B. Uzzi. 2000. *Athena unbound: The advancement of women in science and technology.* Cambridge: Cambridge University Press.

Haraway, D. 1991. *Simians, cyborgs and women: The reivention of women.* New York: Routledge.

Harding, S. 1998. *Is science multi-cultural? Postcolonialisms, feminisms, and epistemologies.* Bloomington: Indiana University Press.

Harding, S. 1996. Rethinking standpoint epistemology: What is 'strong objectivity'? In *Feminism and Science,* eds. E.F. Keller and H. Longino, 235–48. Oxford: Oxford University Press.

Keller, E.F. and H. Longino, eds. 1996. *Feminism and science.* Oxford: Oxford University Press.

Levidow, L. 1999. Whose misunderstanding? *Science as Culture,* 8(2), June: 251–56.

Regal, P.J. 1998. *A brief history of biotechnology risk debates and policies in the United States.* Edmonds: The Edmonds Institute.

Schiebinger, L. 1997. Creating sustainable science. *Osiris 12,* 201–16.

Shiva, V. 1993. *Monocultures of the mind: Perspectives on biodiversity and biotechnology.* London: Zed Books.

Testard, J. and Christian G. 2000. *Au bazar du vivant : Biologie, médecine et bioéthique sous la coupe libérale.* Paris: Édition du Seuil.

SECTION THREE

Weaving New Worlds and Reclaiming Subjugated Knowledges

TIME CAPSULE

We did not learn to love enough
We did not come to see
that we are all sprung
from the roots
of the same tree.

So, in our ignorance,
we turned to hatred and greed,
and fouled our world
and warred with one another
and bred a toxic seed
that could not grow
toward the light.

We choked ourselves
with an insatiable need
for matter,
letting spirit
wither with neglect.
We did not learn respect
for the totality of being.

And now that we are seeing
the end of that unwholesome aspiration
we tried so blindly
to defend,
we leave whomever may come after
a caution and a creed.

Do not do as we have done
Let your lives be filled with love and laughter
and let your consciousness be one.

We pray that you may reach the heights
to which we might have grown
and let our failure
be your stepping stone.

Linda Stitt

12

The Illiteracy of Social Scientists With Respect to Environmental Sustainability[1]

MARGRIT EICHLER

INTRODUCTION

To speak of scholars as illiterate may seem paradoxical. All of us – whether university students, independent researchers, or professors – certainly can "with understanding both read and write a short, simple statement on [our] everyday life," following the early UNESCO definition of literacy. Indeed, all of us will qualify as functionally literate in the sense that we are all able to "engage in all those activities in which literacy is required for effective functioning of [our] group and community and also for enabling [us] to continue to use reading, writing and calculation for [our] own and the community's development" (Lazarus 1985, 3102, see also Quality Standards 1995). At least I hope so.

However, as Willis has noted, "[t]here is no singular history of literacy, nor is there a singular definition of literacy" (Willis 1997, 388). She distinguishes between three broad considerations: literacy as a skill, literacy as school knowledge, and literacy as a social and cultural construct.

In this paper I will use the concept of literacy – and its negative complement, illiteracy – as a metaphor for how the social scientific community deals with matters of environmental sustainability. Using Eisenhart et al.'s definition of scientific and technological literacy, I will examine myself, a sociologist, as an example, to determine to what degree I meet or fail to meet the criteria as specified by this definition. I will then reflect on what would be involved for social scientists to move from illiteracy to literacy in the area of environmental sustainability.

LITERACY WITH RESPECT TO ENVIRONMENTAL SUSTAINABILITY

Today, it is hard to find an educated person who is not aware of global warming and its potentially catastrophic consequences; who does not read almost daily about environmental problems of the gravest dimensions; and who is not cognizant of the fact that private cars contribute greatly to the greenhouse effect, take up an enormous amount of public space, and are a highly inefficient means of moving people by almost any criterion. Most of us are likely to be aware of the dangerous signs of increased immune system deficiencies and to realize that our use of antibiotics has resulted in "superbugs" that are resistant to existing forms of medication. We may be less aware of the many toxins that are daily discharged into our environment because the information is not as easy to come by, but we are certainly aware of the fact that we are exposed to many toxins on a daily basis.

In so far as we are capable of reading about these issues – maybe even talking about them – we are functionally literate in issues of un/sustainability. Some of us are also engaged as citizens in struggling for specific environmental issues. Yet when I open scholarly journals and books in any area that does not specifically specialize in environmental issues, there is little to no indication of a sustainability awareness displayed. This paper concentrates on my own discipline, sociology, but I believe that many of the observations may be applicable to other social science disciplines.

When we turn to more recent definitions of literacy, we find calls for much more wide-ranging skills than those required to qualify as functionally literate according to UNESCO standards. There is, among others, cultural literacy (Zhang et al. 1998), critical literacy (Davies 1997), as well as scientific and technological literacy (Zuzovsky 1997). It is the latter concept that promises to be most helpful in my concern with sustainability. Eisenhart et al. argue that sustainability includes: "a) understanding how science related actions impact the individuals who engage in them; b) understanding the impact of decisions on others, the environment and the future; c) understanding the relevant science content and methods, and d) understanding the advantages and limitations of [the] scientific approach. All these would be indicators of socially responsible scientific literacy" (1996, 284).

Here is a definition of literacy that actually mentions the environment! However, would I qualify as literate under this definition? I have taught graduate courses on ecosociology, ecofeminism, the intersection of equity and sustainability, and the social consequences of reproductive and genetic technologies. I have been struggling for over 10 years to integrate a sustainability perspective into my scholarly work, and have found it exceedingly difficult. I find almost no colleagues in my field who have the slightest interest in this issue. A few students share the concern but are easily discouraged by the lack of conceptual tools that would allow us to deal with issues of environmental unsustainability in a sociologically adequate manner. So how do I personally measure up?

Testing for literacy concerning environmental sustainability

I will now use Eisenhart's criteria to see whether I can consider myself literate in scientific and technological terms and, if so, if I can successfully apply them to declare myself literate in terms of un/sustainability.

a) *Understanding how science-related actions impact the individuals who engage in them*

Here already the first problem emerges. In order to be able to answer whether or not I meet this criterion, I need to understand first what is included under 'science' and who is included under 'individuals who engage in them'. I therefore split the question into these two components and look at them successively.

(1a) What is included as science? The issue here is what do the authors mean when they use the term *science*? The context suggests that "science" refers to the natural sciences, including engineering, but excludes the social sciences and humanities. Even if we accept – for the time being only! – this restriction, there are still multiple forms of knowledge that generate testable, repeatable results that rest on a body of previously accumulated knowledge, but which are excluded from science as conventionally understood. I will here consider only two examples: ethnoscience and women's knowledge.

With respect to ethnoscience, there are, by now, scathing critiques of Western science that demonstrate that it is not the only and certainly not the superior way of knowing the world, and that to privilege this form of knowledge implies a disregard for the ethnoscience upon which a considerable amount of Western science actually rests (for only one example, see Mies and Shiva 1993). For instance, Linnaeus' botanical classification of plants is based on an earlier South Indian classification of pharmaceutical plants (Tsing 1997). This origin tends, however, to be forgotten. Such systemic forgetting is very much in evidence today, as multinational corporations try to patent plants native to southern continents that they have genetically modified. This practice has been characterized as "biopiracy" because it steals the applied knowledge of thousands of years of peasant farmers who have cultivated and experimented with these plants prior to today's efforts to declare them as private property (Shiva 1997). To make such a declaration requires that we treat the preceding work of uncounted people as irrelevant. This is made possible by identifying it as "unscientific."[2]

While modern Western science tends to exclude ethnoscience, it also tends to exclude other forms of knowledge that do not conform to a particular set of rather rigid prescriptions. This has potentially highly detrimental effects for those whose legitimacy is denied if they cannot fit it into these criteria. There is, among other biases, a systemic sexist bias within scientific knowledge.[3] To give just one example, the manner in which occupational health is conventionally studied systematically excludes the knowledge of women who are in turn affected by policies that are based on studies that are fundamentally flawed. This is a systemic bias in the knowledge production and publication process, not an accidental or episodic one (Messing 1998).

The concept of rationality itself, on which so much of Western science rests, has been demonstrated to be flawed because it is premised on a sexist apportioning of human characteristics to women and men (Code 1988).

(2a) Who is included as "individuals who engage in them"? Some of the answers to this question are already contained in the previous comments. In general, ethnoscientists as well as women are excluded from being engaged as actors in the scientific endeavour (as are members of other marginalized groups), notwithstanding the fact that there are women who are scientists themselves. More broadly, do subjects who are exposed to experimental substances – often without their knowledge – count as participating? After all, it is they who experience the effects on their own bodies. In this way, we are all involved in the scientific enterprise, because we are all affected by its results, regardless of whether or not we also engage in scientific experiments as experimenters rather than as subjects.

Do we – and do I – understand the implications of the totality of scientific endeavours, for instance, new weapons' designs, new drugs, new pesticides, new reproductive techniques, genetic research, etc.? Clearly not.

If we take the question in its simplest form and look only at the scientific personnel – and here we would need to include subsidiary personnel, such as office and lab cleaners who may be exposed to toxic substances – do we know the effect of experiments on the experimenters even in this most narrow of senses? How many studies are there that study in detail the effects of, say, animal experiments on the experimenters? Or the effects of studying the human genome on the geneticists who do so? Not many! (And incidentally, it would take social scientists, not natural scientists, to study those effects.)

I, for one, have to conclude sadly that I do not understand how science-related actions impact the individuals who engage in them, whether I interpret this in a broad or narrow sense.

b) *Understanding the impact of decisions on others, the environment, and the future*

This is rather a tall order, especially when it comes to understanding the impact of decisions on *the future*, given that the future holds so many intangibles which are unknowable. With respect to sustainability, often we know only in retrospect that something actually generated a specific set of effects, disturbances in weather patterns being a case in point. More importantly, I am deeply convinced that the scientists involved in conducting experiments do not know what the effects are going to be on other people, let alone the environment or the future. I will provide just three examples of this lack of foreknowledge.

There are a number of horror stories from the pharmaceutical industry where severe problems were generated that became obvious only after a significant amount of time had elapsed (and a significant amount of damage had been done). Examples include the DES tragedy, thalidomide babies, the Dalkon Shield, and silicone breast implants. If we had a different testing mechanism for pharmaceutical products, we would likely have known more about the detrimental effects,[4] but in the absence of testing for long term effects the fact of the matter is that we simply did not know.[5]

Another example is the atomic bomb. Forty years after they had worked on developing the atomic bomb, 70 of the 110 physicists who were involved in the development signed a statement in support of nuclear disarmament (Hynes 1989).

A third example concerns the release of genetically modified organisms (GMOs) into the environment (Kollek 1995; von Weizsacker 1995). We –

humanity – simply do not know what the consequences of this technology will be. This is a new technique being applied widely, for commercial purposes, driven by a profit motive, which has the potential to change the genetic structure of organisms other than the ones experimented upon.

To me this history suggests very strongly that in all instances we should proceed much more slowly – if at all. We should test drugs and other pharmaceutical products for much longer before they are prescribed on a mass basis, and if we genetically engineer plants and animals at all (which should involve, I submit, a public debate that has not yet happened in North America)[6] – it should be done much much more slowly than it is currently. As far as weapons of mass destruction are concerned, many people would agree that they should not be produced at all.

While I can, therefore, forecast some negative effects of *some* activities, such as the continued release of greenhouse gases into the atmosphere, I am certainly unable to understand how specific scientific actions in general will affect anybody or anything. On this criterion as well, I must confess that I am illiterate.

c) *Understanding the relevant science content and methods*

In this instance, I blame my high school education, which prepared me very badly indeed to understand scientific content. While I have some grasp of what is usually referred to as "the" scientific method, I am woefully ignorant of the content of the natural sciences. Another issue becomes relevant here: the assumption seems to be that there is only one scientific method – an assumption I do not share – and furthermore that this involves the only relevant way of knowing. While "the" scientific method is, indeed, *one* powerful and important means of understanding some aspects of the world, it also misses many other aspects, particularly the social conditions that shape the way specific problems are posed (Harding 1992).

It is also interesting in this context to ask ourselves why it is seen as important for non-natural scientists to be familiar with the methods and content used by natural scientists while there is no such requirement laid on natural scientists to understand social scientists and humanists. It seems to me that we, too, have something to offer the world that is of real import.

Be that as it may, when it comes to the content of natural science, once again I must confess my illiteracy.

d) *Understanding the advantages and limitations of the scientific approach*

This is the one point where I feel a bit more confident. I feel that I do understand the advantages of the scientific method, and I am very much aware of its limitations. This would be the one aspect of the overall set of requirements where I would consider myself functionally literate.

HOW TO UNDERSTAND MY ILLITERACY

Sadly, meeting one out of four criteria does not suffice to make one able to declare oneself as possessing socially responsible scientific literacy. This disappointing situation may be due partly to my deficient high school education, my failure to make up for it in the many intervening years, or the fact that I am simply a particularly obtuse learner.

However, attributing my illiteracy only to my own personal limitations "[f]ails to acknowledge the specific contexts and constraints of literacy" (Willis 1997, 5) and fails to acknowledge the social and cultural construction of my illiteracy. If I were the only person who failed the test of literacy as described above, I would feel fairly secure in the knowledge that others are simply smarter than I am and that I can, therefore, rest secure in their literacy and the judgments that flow from it. However, as I have argued above, some of the issues are in principle unknowable, at least under current circumstances.

At this point we return to where we started. When it comes to environmental sustainability, all the colleagues I have so far sampled admit that our current way of life is unsustainable, but the vast majority draw no consequences *in terms of their scholarly work* from this. They may, of course, be active in various ways in their role as citizens. Nonetheless, in so far as the issue of whether conditions on the globe will be able to continue to support human life within the next 50 years could be expected to be of *some* interest and relevance for anyone who is paid for thinking about the world, the absence of a widespread concern about unsustainability in our scholarly work points to a structural barrier.

Literacy research demonstrates that historically, access to literacy skills was restricted on the basis of gender, race, ethnicity, language, and geographical location. "[D]efinitions and purposes of literacy have always been closely connected with the ideological, political, social, racial, and economic goals of nations" (Willis 1997, 389).

We therefore need to ask in whose interest is it to keep us illiterate with respect to sustainability? A reasonable guess would be to look towards those who profit most from the existing economic system. We are, by now, used to hearing that the richest 20 percent of the world's population utilize over 80 percent of the world's resources, and that the inequality is increasing (Asian NGO Coalition et al. 1993 and 1994). All of us who belong to the middle class in highly industrialized countries belong in the category of those of profit – at least in the short run – from the wasteful and unsustainable organization of the world. While this is an important point, never to be forgotten, it nevertheless hides the fact that there is a category of superprofiteers in comparison with whom even our benefits pale. The combined assets of the three richest *individuals* equal the GDP of the 48 poorest nations. A meagre four percent of the combined wealth of the 225 richest people in the world would suffice to pay for achieving and maintaining universal education and basic health care for everyone, reproductive health care for all women, and safe water and sanitation for all (United Nations 1998, 30). One has to struggle to imagine this: four percent of the combined riches of 225 people would be enough to solve some of the most significant social problems in our world.

With such an almost unimaginable concentration of riches in a few hands, which in Canada goes along with a similar concentration of the news media which are – or are supposed to be – our primary means of information about current affairs, we are being kept systematically illiterate with respect to sustainable ways of organizing ourselves. To take an example at random, a lead article on page 1 of the *Globe and Mail* alerts us in big bold letters: "The Economy is lagging, Canada warned" (Little 1998, A1). It then goes on to inform us that the "OECD foresees a major tumble in our relative standard of living" (A1). "The OECD projection implies a bleak future for average Canadians. After falling between 1989 and 1992, GDP per capita has increased in each of the past six years. Even so, personal income per head has fallen steadily since 1990, especially after-tax income" (A1). The solution to Canada's supposed woes? "[C]ut taxes, put more people and capital to work, wipe out interprovincial barriers to economic activity, and become more productive" (Little 1998, A1).

What is the relevance of this prescription for environmental sustainability? Leaving aside the issue of social equity – the per head personal income decline mentioned is in no way equally distributed (I have received a tax cut while welfare recipients have suffered a substantial cut to their income and cutting taxes even more will be good for me but not for the homeless), we need to "wipe out interprovincial barriers to economic activity and become more productive." Unsaid, but very much a part of the agenda, is to wipe out trade barriers everywhere, and allow capital to move unfettered by any ties to localities, thus increasing existing inequalities and decreasing further any responsibility for the environmental deterioration that may be caused (Asian NGO Coalition, 1993 and 1994).

Finally, what will increase productivity? Anything that involves money – even gambling by opening even more casinos, clear-cutting old forest growths, building more prisons, hiring more prison guards, because they are paid in money and thus the GDP is increased by the amount of their salary. Of course, we all know that environmental disasters are also wonderful for the balance sheet – all that economic activity that follows in their wake! (Waring 1988) Producing nuclear reactors – one of Canada's exports – counts certainly as being "productive" and will "put ... people and capital to work." None of these things are likely to increase our environmental sustainability. They are, however, likely to profit the rich (the middle class in the highly industrialized countries) and particularly the superrich.

Nevertheless, I do not think that we can put all the blame on the super-profiteers and on neoliberal and neoconservative economics (the distinction between which continues to elude me, in practice as well as in theory). The neglect of the importance of natural environmental factors within sociology goes back to the very beginning of the discipline. There are various accounts as to *why* sociology fails to take the natural environment seriously. They range from identifying our ideology as one of exuberance and human exemptionalism (Catton and Dunlap 1980), over our tendency to look for the *meaning* of matters rather than their direct effects (Hannigan 1995; Newby 1991) to the perceived need to differentiate ourselves from the natural sciences.

Looking at the presidential addresses of the presidents of the American Sociological Association – a group of sociologists who are by definition highly influential beyond their own country, and who represent a wide diversity of philosophical, epistemological, and methodological approaches and an equally wide range of substantive interests – we find that over a hundred-year span not a single one of them foregrounded the need to make our society environmentally sustainable (Eichler 1998). There are, of course, some sociologists worldwide who are very much concerned with this issue, but they do not at present represent the mainstream. Their work is not seen as foundational for everything else – which, given the absolute necessity of functioning life support systems for every human activity, including the pursuit of sociology or any other social science – I submit it should be.

The issue is not simply one of including the natural environment in our considerations, but of recognizing explicitly that our survival as a human species is threatened by our unsustainable way of organizing ourselves. This requires, in my mind, that we acknowledge this fact and make it the touchstone against which all other issues are assessed. This is not a suggestion that sits easily with most social science scholars.

My own work has been oriented around feminist issues since I started teaching a long time ago, and I continue to see the value in and need for a feminist perspective. Feminist theory is one place that actually *has* generated a new approach that includes issues of sustainability, namely ecofeminism, and this provides a starting place to find concepts and theoretical approaches to make the necessary connections.[7] While other equity-seeking efforts (anti-racist thinking, sexual diversity approaches, and other attempts to include hitherto excluded groups) clearly make important contributions to our understanding of social inequities and hopefully pave the way towards moves for greater social equity, not many of them are also concerned with issues of environmental sustainability. There are some exceptions, such as the literature on environmental racism (See Bryant 1995; Bullard 1994; and several articles in Johnston 1994).

MOVING TOWARDS LITERACY

What then, would constitute literacy with respect to sustainability? I suggest the following criteria:

a) a basic acknowledgement *in our scholarly work* that all human efforts are contingent on the existence of the earth's life support systems, that these life support systems are currently threatened through human activities, and that this must be reversed if we wish to survive as a species;
b) a critical exploration of how our own subject matter and approach contributes to the maintenance of an unsustainable society, or decreases or increases environmental unsustainability; and
c) the development, within our scholarly work, of pointers towards decreasing environmental unsustainability.

The focus is on *unsustainability* because this is a lot easier to recognize than what is needed for sustainability. For instance, we can all agree that overpackaging, non-returnable pop containers, greenhouse gas emissions, and so on constitute unsustainable practices. There will be much more resistance in classifying private cars as an unsustainable form of mass transportation, but the evidence is nevertheless very clear on this issue (Zielinski and Laird 1995; Zuckerman 1991). Focusing on eliminating unsustainable practices is more down-to-earth than trying to agree on what the perfect sustainable solution would be. Besides, by identifying some aspect of our current way of life, and agreeing on eliminating or drastically reducing it, we free the mind (and, with proper taxation, the market) for a multitude of alternative solutions, rather than just one. (For examples of positive solutions, see Nozick 1992; Korten 1996; O'Hara 1993; and Roberts and Brandum 1995).

We may still ask ourselves how useful it is to spend time on issues of sustainability. In my case, assuming that my subject matter is pretty far removed from it, the effort is herculean, and it is likely not to have much effect anyhow, given the immensity of the various combined problems. Here I find it useful to turn to literature as a source of inspiration. In her visionary book *The Fifth Sacred Thing*, Starhawk describes a post-ecological disaster world in which one city has managed to rebuild an ecologically sane and socially equitable society. At the point at which they are about to be invaded by the military of the neighbouring totalitarian state, the main character, Maya, reflects: "The only war that counts is the war against the imagination." ... All war is first waged in the imagination, first conducted to limit our dreams and visions, to make us accept within ourselves its terms, to believe that our only choices are those that it lays before us. If we let the terms of force describe the terrain of our battle, we will lose. But if we hold to the power of our visions, our heartbeats, our imagination, we can fight on our own turf, which is the landscape of consciousness. There, the enemy cannot help but transform (Starhawk, 1993, 238).

If nothing else, we can chip away in our own little way at the prevailing image that our unsustainable way of life is unalterable, and that there is no alternative to the neoliberal/neoconservative economic agenda. We all can and need to make a contribution, however modest, to making our society more sustainable. The call for a literacy campaign to eliminate unsustainable ways of thinking and acting therefore needs to start with ourselves.

NOTES

1 I would like to thank Diana Gustafson for research assistance for this paper. This is a revised version of a paper presented at the University College Symposium on Literacy, Toronto, January 1999.
2 This is quite apart from the fact that many consider the patenting of any life form for any purpose to be morally abhorrent.

3 Sexism can take a multiplicity of forms, and is here used as an overarching concept, cf. Eichler, 1991 and 1997.
4 Baraldi (1994, 72) describes the process as follows for Canada: "the long-term effects of a new pharmaceutical product have not been evaluated at the time of marketing. At best, the product may have been administered to 3000 patients, a limited sample in which even an undesirable reaction with high incidence would have little chance of being noticed ... For instance, a retrospective analysis of products withdrawn from the market in recent years showed that several hundred thousand people had to be exposed to the effects of a product before its harmfulness, on the short or medium term, was recognized ... If this is true of short- and medium-term effects, we may assume that much more time and the exposure of many more persons are needed before the long-term effects of a product will be identified."
5 Some people would argue that there were, in fact, scientists who did know about the potential adverse effects but failed to act on this knowledge. Even if this were the case, I believe that there were many ethical doctors who prescribed these treatments because they were ignorant of the effects.
6 The public debate on this issue is much more lively in Europe. For an example, see *The Ecologist*, 1998, 28 (5).
7 The literature on this issue is voluminous. Gaard (1998, 11–52) provides an excellent introduction into it by tracing the various contributing rivers of thought.

REFERENCES

Asian NGO Coalition, IRED Asia and the People-Centred Development Forum. 1993. Economy, ecology and spirituality: Toward a theory and practice of sustainability (Part I). *Development (Journal of the Society for International Development)*, #4:74–80.
– 1994. Economy, ecology and spirituality: Toward a theory and practice of sustainability (Part II). *Development (Journal of the Society for International Development)* 4, 67–72.
Baraldi, R. 1994. The evaluation of pharmaceutical products: Problems of phase IV. In *Misconceptions. The social construction of choice and the new reproductive and genetic technologies Vol. 2*, eds. G. Basen, M. Eichler and A. Lippman, 71–81. Prescott: Voyageur Publishing.
Bryant, B. 1995. Issues and potential policies and solutions for environmental justice: An overview. In *Environmental Justice. Issues, Policies and Solutions*, ed. B. Bryant, 8–34. Washington: Island Press.
Bullard, R. 1994. Environmental racism and the environmental justice movement.

In *Ecology*, ed. C. Merchant, 254–65. New Jersey: Humanities Press.

Catton, W. R. Jr. and R. E. Dunlap. 1980. A new ecological paradigm for post-exuberant sociology. *American Behavioral Scientist* 24(1), 15–47.

Code, L. 1988. Credibility: A double standard. In *Feminist perspectives. Philosophical essays on method and morals*, eds. L. Code, S. Mullett, and C. Overall, 64–88. Toronto: University of Toronto Press.

Davies, B. 1997. Constructive and Deconstructive Masculinities. *Gender and Education* 9(1), 9–30.

The Ecologist. 1998. The Monsanto files. Can we survive genetic engineering? 28 (5), Sept/Oct.

Eichler, M. 1991. *Nonsexist research methods. A practical guide*. New York: Routledge.

– 1997. Feminist methodology. *Current Sociology* 45(2), 9–36.

– Towards a more inclusive sociology. *Current Sociology* 46(2), 5–28.

Eisenhart, M., E. Finkel and S. Marion. 1996. Creating the conditions for scientific literacy: A re-examination. *American Educational Research Journal* 33(2), 261–95.

Gaard, G. 1998. *Ecological politics. Ecofeminists and the greens*. Philadelphia: Temple University Press.

Hannigan, J.A. 1995. *Environmental sociology. A social constructionist perspective*. New York: Routledge.

Harding, S. 1992. The instability of the analytical categories of feminist theory. In *Knowing women: Feminism and knowledge*, eds. H. Crowley and S.Himmelweit, 338–54. Cambridge: Polity Press.

Hynes, H.P. 1989. *The recurring silent spring*. New York: Pergamon Press.

Johnston, B.R. 1994. *Who pays the price? The sociocultural context of environmental crisis*. Washington: Island Press.

Kollek, R. 1995. The limits of experimental knowledge: A feminist perspective on the ecological risks of genetic engineering, In *A feminist and ecological reader on biotechnology and biopolitics*, eds. V.Shiva and I. Moser, 95–111. London: Zed Books.

Korten, D.C. 1996. *When corporations rule the world*. West Harford: Kumarian Press.

Lazarus, R. 1985. Literacy and numeracy: Policies. In *International encyclopedia of education*, ed. T.Husen and T.N. Postlethwaite, 3102–205. Oxford: Pergamon Press.

Little, B. 1998. Economy is lagging, Canada warned, *Globe and Mail*, Dec. 3, pp. A1 and A11.

Messing, K. 1998. *One-eyed science: Occupational health and women workers*. Philadelphia: Temple University Press.

Mies, M. and V. Shiva. 1993. *Ecofeminism*. London: Zed Books and Halifax: Fernwood.

Newby, H. 1991. *One world, two cultures: Sociology and the environment*. (Lecture given to mark the 40th anniversary of the founding of the British Sociological Association). Network 50, (February).

Nozick, M. 1992. *No place like home. Building sustainable communities*. Ottawa: Canadian Council on Social Development.

O'Hara, B. 1993. *Working harder isn't working. How we can save the environment, the economy, and our sanity by working less and enjoying life more*. Vancouver: New Start Books.

Quality standards for adult literacy. A practitioner's guide. 1995. (The accountability framework for the adult literacy education system and core quality standards for programs). Literacy Section. Learning and Employment Preparation Branch. Ontario Training and Adjustment Board.

Roberts, W. and S.B. 1995. *Get a life! How to make a good buck, dance around the dinosaurs and save the world while you're at it*. Toronto: Get a Life Publishing House.

Shiva, V. 1997. *Biopiracy. The plunder of nature and knowledge*. Toronto: between the lines.

Starhawk. 1993. *The fifth sacred thing*. New York: Bantam.

Tsing, A.L. 1997. Transitions as translations. In *Transitions, Environments, Translations. Feminisms in International Politics*, eds. J.W. Scott, C. Kaplan, D. Keates, 253–272. New York: Routledge.

United Nations Development Programme. 1998. *Human development report 1998*. New York: Oxford University Press.

Waring, M. 1988. *If women counted. A new feminist economics*. San Francisco: Harper and Row.

Weizsacker, C. von. 1995. Error-friendliness and the evolutionary impact of deliberate release of GMOs. In *Biopolitics. A feminist and ecological reader on biotechnology*. V. Shiva and I. Moser, eds., 112–20. London: Zed Books.

Willis, A.I. 1997. Focus on research: Historical considerations. *Language Arts* 74(5) Sept, 387–97.

Zhang, C., L.O. Ollila and C.B. Harvey. 1998. Chinese parents' perceptions of their children's literacy and schooling in Canada. *Canadian Journal of Education* 23(2), 182–90.

Zielinski, S. and G. Laird, (eds.) 1995. *Beyond the car*. Toronto: Steel Rail Publ./Transportation Options.

Zuckermann, W. 1991. *End of the road. The world car crisis and how we can solve it*. Cambridge: Lutterworth Press and Post Mills, Vermont: Chelsea Green.

Zuzovsky, R. 1997. *Studies in Educational Evaluation* 23(3), 231–56.

13

Teaching in Engineering As If the World Mattered

MONIQUE FRIZE

INTRODUCTION

Some of the suggestions discussed in this paper have evolved from the report prepared by the Canadian Committee on Women in Engineering (CCWE) released in 1992. The rationale behind the formation of the CCWE was the predicted shortage of female engineers. The causes of the anticipated shortage were cited as: decreasing enrolments in engineering programs; the dwindling number of engineers immigrating to Canada; and future economic growth. In addition, the tragedy at École Polytechnique in Montreal jolted engineers and non-engineers alike into an open discussion of the issues that limit the representation of women in this profession. The process undertaken by the CCWE was consultative and took the form of public forums, submission of written briefs, a national conference and, in September 1991, a meeting of stakeholders who further developed the CCWE's recommendations. The CCWE report (1992) addressed a wide spectrum of conditions affecting the level of participation and the status of women in the profession. Although some of the twenty-nine recommendations have been implemented, especially at the pre-university level, many of the strategies proposed have yet to be implemented by universities, workplaces, and the profession itself. This paper focuses primarily on strategies that could improve the attraction and retention of women into engineering undergraduate programs and on how to make students more aware of their future responsibilities to society and to the environment.

THE CHANGING PROFILE OF ENGINEERING

In Canada, the demographics of the student population in Engineering are changing. The average proportion of women in engineering undergraduate programs in 1985 was 11 percent; by 1991, it was 15 percent, 19 percent in 1995, and 21 percent in 2001. Enrolments of full-time women students in master's programs were 10 percent in 1989, 20 percent in 1995, and 24 percent in 2001. In doctoral programs, 6 percent were women in 1989, 13 percent in 1995, and 17 percent in 2001. Women comprised 2.2 percent of engineering faculty in 1991, and 7.6 percent in 2001 (CCPE).

In 2000 and 2001, the number of men enrolled went up by one to two percent in many of Ontario's engineering schools, while the number of women stayed approximately the same. This meant a corresponding drop in the absolute proportion of women. Engineering programs are already male-dominated, so we have to work to ensure that this decrease doesn't continue, in order to maintain some gender diversity in our student population.

If all students are to develop their full potential, the changing profile of engineering students must be matched with appropriate adjustments to the curriculum, culture, and environment to suit the learning style and needs of the various groups. It is especially important that women not just emulate the male model, but be allowed to incorporate their experiences and perspectives into their study and into their work.

The concept of gender differences in learning has been studied most extensively among pre-university students. For example, Pat Rogers at York University (1988) has studied gender differences in learning mathematics. This type of research is very valuable in identifying how teachers can optimize learning and the development of problem-solving skills in all their students. Belenky *et al.* (1986) discuss women's ways of learning; Sue Rosser (1990; 1997) suggests ways to make science friendly to female students. Tobias (1990) presents several case studies to show how the traditional teaching style in science courses turns away bright male and female students who might have responded better to a less traditional environment. More recent discussions of curriculum, teaching style, and the culture in engineering classrooms have been published, amongst which are the writings of Tonso (1997) and several papers presented at the International Conference of Women Engineers and Scientists (ICWES12 2002)

In addition to course content, teaching methods also need to be modified to reach those who prefer relational as opposed to rule-based pedagogy. Pat Rogers (1988) argues that the expository mode of presentation is androcentric. In contrast, relational thinkers (this applies to many women and to some men) prefer working together in small groups.

In middle-class, White North America, Carol Gilligan's (1982) work suggests that males are encouraged to challenge authority, to hold the floor and express their own views while women are socialized to subordinate their views to other authorities and not to develop a sense of their own autonomy. Following this argument, these stereotyped and learned attitudes tend to favour men in a masculine learning process such as that found in engineering. Professors must develop a sensitivity toward these issues and develop ways to integrate different ethical and cognitive styles into the learning process. Classroom situations must be developed to promote women engineering students' confidence and to help them overcome some of the societal barriers they face. In the past the erosion of women's self-esteem during their high school years has been well documented and must urgently be addressed, as many women later enter post-secondary education with a serious handicap created by the inequities of the elementary and secondary educational system (Greenberg-Lake 1991).

Some interesting studies have been done on gender differences in learning at the post-secondary level and on how students perceive the quality of their student life (Light 1990; Drolet and Turgeon 1991). Gender appears to have a substantial influence on academic performance and on the retention of students. If engineering continues to be an education by and for men, the number of women students may not rise beyond the current level or may even decline.

Since engineering has been a field with little feminist input, it can be assumed that the engineering environment, knowledge construction, and teaching are androcentric in perspective. An important lack in previous generations of engineering graduates was people skills. In Canada, the curriculum includes complementary courses in the arts and humanities, but frequently students choose courses in which they can get good grades rather than courses that would provide the most knowledge and experience in verbal, writing, and interpersonal skills. Although guidelines created by the Canadian Engineering Accreditation Board (CEAB of CCPE) recommend that the curriculum include information on the impact of engineering on society and on the environment, few engineering schools have incorporated this into their programs. One exception is McMaster University which has been offering a full undergraduate degree in Engineering and Society for some years. It is critical to include these considerations in all engineering programs and make students understand their future responsibilities with regard to these issues.

Vickers et al. (1995) found that three-quarters of the young women (enrolled in their first year of science or engineering) who participated in their study said that helping society was an important factor in choosing a career, whereas only half of the young men thought that this was important. A study by Pomfret and Gilbert (1990) reported that women tend to select courses and careers that will allow them to help others and society. Another study shows that women engineering students want to work on humanitarian concerns in a professional environment. Drolet and Turgeon (1991) state, regarding women, that: "Paradoxically, their desire to contribute to humanity is not encouraged [in engineering] and becomes an obstacle to their progress." Michaela Serra, a computer science professor said in her brief to the CCWE (1992), "The idea of establishing or emphasizing more socially beneficial programs in engineering may be inspired by needs of women in engineering, but can profit everybody: men and women in engineering, and the population at large."

Are we producing the kind of engineer who will be needed in the twenty-first century? We do not want to prepare engineers for corporate practices that destroy the environment in the highly competitive cut-throat process of globalization. Instead, we would like to prepare engineers to prioritize local and environmentally friendly initiatives for solving technological problems within communities (like those seen in the film, *Asking Different Questions: Women and Science* featuring Ursula Franklin, Rosalind Cairncross, Karen Messing, and Peggy Tripp). As for the types of corporations that hire engineers, we hope that they operate "as if the world mattered." Some organizations have reshaped their mission statements and are adopting a new style of management and a new corporate culture, showing that they are conscious of their societal responsibility. Companies ignoring these considerations are bound to face increasing opposition and may even disappear in the early part of this new century.

At the CCWE forum in Montreal in 1991, Micheline Bouchard, vice-president of a middle-sized company and incoming president of the Canadian Council of Professional Engineers, said: "The [engineering] profession will have to adopt more values that are said to be feminine. Already engineers' social awareness is allowing the emergence of new values, such as protection of the environment, health, and occupational safety. Engineers are regularly called upon to work in multidisciplinary teams. They work with economists, biologists, lawyers, and leaders of labour movements. They can no longer sell products and services based only on technical know-how. They must understand the needs of their environment and explain how they will meet those needs."

An increasing number of employers want their engineers to have highly developed communication skills and a leadership style that empowers employees and integrates diversity of the workforce. They want well-developed interpersonal skills and a strong ethic. Many of these qualities have traditionally been regarded as feminine and non-Western attributes. Adding these qualities to the skills of graduates will help create more civilized and well-rounded engineers (Florman 1987).

THE CULTURE IN ENGINEERING IN CANADA

The culture of a profession is sculpted by the values, norms, styles of discourse, and relations of power that lie behind it. This includes rituals, forms of talk, conformity of behaviour and interaction, mode of dress, and the image projected by the individuals who belong to the group. In university programs, faculty and students usually share these values. Hacker (1981) describes the culture of engineering as an environment that stresses the importance of technology over personal relationships, formal abstract knowledge over inexact humanistic knowledge, and male attributes over female ones. This has been corroborated by other authors (Robinson and McIlree 1991) who add: "To look like an engineer, talk like an engineer, act like an engineer, in most places, means to look, talk, and act male." Sorensen (1992) observes that these androcentric characteristics also define the high-tech world of work and that this perhaps partly explains why women are not present in large numbers in the engineering industry. The difficulty for women who wish to belong to this group is to maintain their femininity while projecting an image of engineering competence. According to Tonso (1997), this is a difficult task. Speaking of her own experience as an engineering student and of her current research on gender issues in engineering education, she says:

When I entered engineering over 25 years ago, becoming an engineer seemed to imply that I could go out into the world and assume male duties. That presumption comes from an earlier day and age. Times have changed and we must now understand that being forced to adopt, in an unexamined way, these culturally and historically taken-for-granted ways of construing professional identity is a violence against women that not only denies our existence, but also denigrates the importance and honor of women's work. To expect that women "check their identities at the door" constrains our participation – makes us silent and invisible.

Today engineering is still perceived as a career for men, and this is true in all parts of the world. Although women have been involved in technology and science in all historical periods, these women were exceptions and not the rule. When the settings for scientific and technological activities were rather informal, as in salons, monasteries, or in people's own homes, it was easier for women to participate in these activities than when the process was formalized through the creation of the Royal Societies, the Guilds, the Professional Bodies, and formal educational curricula in the nineteenth and twentieth centuries.

Professional programs such as engineering tend to have similar curricula, partly due to the fact that Anglo-American influence is felt all over the world just as it was in colonial times. That is, universities everywhere tend to model their engineering curricula after the dominant schools. This is not to say that in engineering schools no progress has been made in the past 10 years, especially where blatant sexist behaviour is concerned. However, some student publications, from time to time, still include sexist and inappropriate material. But discrimination still exists; it has merely gone underground, becoming more subtle. Several female students state that they can't exactly say what makes them feel uncomfortable, but that they are sometimes made to feel they do not belong, sometimes by professors, at other times by male peers. This attitude has been called "the chilly climate." This expression refers to the subtle exclusion of women by language use, increased attention to the contributions of men in the class, and a higher assessment of work assignments completed by men (Sandler, 1986). Pat Rogers (1988) quotes the Ehrhart and Sandler study (1987): "Behaviours and attitudes that express different expectations for women or single them out or ignore them because of their sex ... can have a profound negative impact on women's academic and career development." The most commonly cited complaint heard by the CCWE (1992) from women engineers was that they felt that they were not taken seriously. In a written brief, a woman said: "I remember one incident where a male peer said he could never work with a woman and take her seriously. This was quite a shock, as the same individual certainly took me seriously enough to borrow my assignments!" In other cases, the anger goes underground and shows up in blank faces and chilly attitudes when gender issues are discussed in a classroom setting.

The CCWE report (1992) stated that a critical step in increasing the participation of women in engineering is to hire more women in engineering Faculties. As stated previously, the level of female faculty in Canadian Engineering Schools/Faculties was close to eight percent in 2001. These few women are undoubtedly over-extended with counselling, committees, teaching, and research. Increasing the number of women faculty is a key step in creating a more comfortable environment and a gender-balanced curriculum. These women can be role models for younger women who may be considering such a career. It is also important that male and female students come to see women in engineering as normal and beneficial to the profession. While the pool of female doctoral candidates in engineering remains small (around 17 percent), proactive methods to recruit women professors must be considered. The creation by NSERC of Women's Faculty Awards (1990 to 1995) and more recently, the University Faculty Awards, are important steps in increasing the participation of highly qualified women in academe, specifically in the disciplines of science and engineering where their presence is very low (NSERC, 1996). Other short-term suggestions are hiring women from industry to teach on a part-time basis; hiring women with master's degrees as instructors; or hiring women with master's degrees as assistant professors and providing financial assistance while they complete their doctorates. Laval University did this successfully in the 1990s. Appointing women as adjunct professors is another way to increase the presence of women, although this is less effective than providing them with tenure-track positions. In Europe, a recent report decried the paucity of women professors in all major nations of the EU and major efforts and funding are being deployed to increase the number of women faculty in their science and engineering faculties (ETAN report, 2000).

TEACHING ENGINEERING AS IF THE WORLD MATTERED – DEVELOPING A SOCIAL CONSCIENCE IN OUR ENGINEERS

Cultural changes do not occur overnight, particularly when a profession has a long history and a conservative approach to change. Engineering is such a profession, and until cultural shifts occur at all levels, that is, with the professors, the students, the alumni/ae, and with those who currently govern and regulate the profession, other more immediate steps are needed. These are necessary if we wish to improve the positive impact of engineering on our world, minimize its negative effects, and improve the climate for women and for other underrepresented groups within the profession.

First, there are very few programs that have integrated considerations of social context into the engineering curriculum (Vanderburg 1990). This concept must spread to all engineering faculties in Canada if a substantial connection between engineering and society is to be made. Faculties of engineering should also encourage engineering students to take courses in which engineering, humanities, and social science students interact. Interaction among students of various disciplines will allow engineering students to develop an understanding of non-engineering students and will give non-engineering students the opportunity to become more comfortable with technical specialists. Engineering students should be encouraged to take courses that highlight the contributions of women to the sciences and the arts, as this would promote an acceptance of the equality of women in society and in the workplace.

A second approach tested over a decade (Frize 1996) has been to add a core component to engineering education that attempts to develop the social conscience of students. In most Canadian engineering schools/faculties, a course on professional practice is mandatory and must be successfully completed by the student in order to graduate. A survey conducted by the author in the summer of 2000 found that the course content and the manner in which it is taught varies from one university to the other. The professional practice course that exists at many engineering schools focuses primarily on the engineer's responsibility to clients and to the public. But codes of ethics should also apply to the way engineers treat their colleagues and their employees. Faculties of engineering should explain to students the meaning of fair hiring and promotion practices and the need to eliminate sexual harassment and to be respectful of all co-workers. Even the manner of evaluating the students in professional practice courses varies from a simple pass/fail system through the assessment of a logbook in which the lecture notes are compiled, to a full assessment with logbooks, two essays, and a final exam worth 50 percent of the final mark. In the first model, it is not surprising that students frequently do not take the course seriously.

Having given some thought to these challenges, the author began to see ways to provide equity-relevant skills within the existing curriculum, by changing some of the approaches to teaching. One of the most obvious places to begin was the professional practice course. What was really distressing was the exclusive use of male language by some of the women in my class, even when referring to their own experiences. This showed how engineering students (including women) have been conditioned to see engineering and engineers exclusively in male terms. This is a systemic problem and it will only disappear by continuing to stress the use of inclusive language in the classroom.

In 1990 the author restructured the course in Electrical Engineering at the University of New Brunswick (and at Carleton University in 1997) by adding a discussion of gender issues; addressing why the enrolment of women in engineering was still low compared to faculties of law, medicine and business; discussing workplace issues such as employment equity and sexual harassment; and stressing the importance of acquiring people skills. These lectures were also provided to other sections of the course in mechanical and in civil engineering.

Although professional practice course content may vary slightly among instructors across the country, we could (and should) develop a "guideline" to ensure that a number of these issues are covered. People skills can be developed by assigning writing tasks and verbal presentations. In the new course design, approximately half of the guest speakers were professional women, several of them engineers. Invited speakers have been asked to use gender-inclusive language and examples that include women. Topics added to the course include:

Societal Impact

A history professor was invited to discuss the social impact of technology with the students. This talk was followed by the writing of an essay in which the students were asked to invent an engineering job that they have enjoyed for about five years and to discuss how their work impacts society both in a positive and a negative manner. Most students understood the objective and did a very interesting paper, although they struggled more to find the negative aspects than the positive ones. They often concluded in this way: "Although there are some negative impacts of my work on society, the positive impacts far outweigh the negative ones." That is a normal reaction, since they chose to be engineers and thus to be creators of new technologies. But students frequently admitted that the essay helped them to think about these concepts.

Environmental Concerns

An environmental engineer discussed how the environment has to be a consideration for any engineering project and encourages the students to respect the public, to treat people who are not engineers with respect and to learn how to address them in terms that they understand, instead of the usual professional jargon that we engineers are so prone to use in conversations amongst ourselves. Principles of conservation and sustainable development were presented.

Engineering Workplace and Gender Issues

There was a discussion of current obstacles for women in the profession and strategies to achieve gender balance. The concept of employment equity and the definition of sexual harassment, both in university and in the workplace, were presented. Future engineers also need to know that many employers expect men to be able to work successfully with and for women, and that they have adopted workplace policies that prohibit homophobia, racism, and sexism.

Ethical Theories and Ethical Decision-Making

Four ethical theories were described as well as the process of ethical decision-making when faced with moral dilemmas in engineering professional work. Several case studies were assigned to the students as one of the two main assignments of the course.

Issue of Privacy

Our privacy is being invaded more and more today, even by governments who purport to be democratic. Examples are the cameras installed in many public places in the UK; The ECHELON system that detects the emails and telephone conversations of private citizens (UK and USA, with participation from several other Western countries); and the global file that was being compiled by HRDC (Human Resources and Development Canada) on Canadians in the 1990s. It is essential in this new century to raise the awareness of students about the issue of privacy, particularly for those who are the future designers of information technologies.

Other Topics

Most professional practice courses in Canada discuss the role of the provincial and territorial professional engineering associations; that of the Canadian Council of Professional Engineers; and of the scientific associations such as IEEE (Institute of Electrical and Electronic Engineers); the importance of ensuring a safe working environment as per the policies and laws on Occupational Health and Safety; and other topics related directly to the specific discipline of engineering. The issues of liability and how to minimize this risk, how to plan a career, the importance of continuing education, and the legal aspects of engineering such as contract law and intellectual property laws were also covered.

The women and men guest speakers demonstrated their competence to the students and shared interesting experiences with the class. This has had a positive impact on the attitude of the students. The students were also asked to use gender-inclusive language in the classroom, in their essays, and in their case study analyses. The majority of them do so in the classroom, as well as on their examinations.

After several years of observing this new course content and structure, results have been generally positive. Most of the older students (those in fourth year) seem to understand the points made and their logbooks reflect this. However, younger students (those in third year) respond differently and a few think it is a waste of their time to be presented with ethics and gender material. The writing skills of most students improve during the term. Their greater understanding of the new workplace environment seems to diminish the fears and anxieties they may have had about issues such as equity and harassment. Being well prepared, they are better able to handle stress and the unexpected. Many of the myths held by the students at the beginning of the course were replaced by facts and real-life examples. The role models who visited the class provided real-life reinforcement of the main concepts through their own messages.

SUGGESTIONS FOR TEACHING TECHNICAL COURSES AS IF THE WORLD MATTERED

Regular technical courses should include references to societal impact to make them more relevant to the students. This would go a long way to sensitize students to the real-life societal impact of their future engineering work.

A BETTER CLIMATE FOR WOMEN STUDENTS

The number of women studying engineering may not increase substantially until the curriculum is made more gender sensitive, and more relevant to concerns of equity and sustainability. Courses that emphasize the beneficial applications of technology and engineering will increase the appeal of engineering programs to women. Examples include solving environmental problems; inventing tools and equipment for the disabled; designing and constructing safe highways; and developing user-friendly technologies for use in health care institutions and for home-care programs. All engineering students need to understand the complex nature of technological decision-making and to realize that the best solutions are based on a blend of technological and political considerations and the values of the society within which the projects take place.

The average attrition rate of women in engineering has been around 40 percent (Pomfret and Gilbert 1991). Efforts to attract women are wasted if women cannot progress academically and in the workplace with success and confidence. A study (Drolet and Turgeon 1991) concluded that young women have particular characteristics that make it more difficult for them to persist in their studies when difficulties are encountered. The study found that the women were more perfectionistic, were harder on themselves than men, and attributed results to effort rather than talent. They also tended to prefer collaboration to competition and to place more importance on a supportive environment. To the extent that the results of this study can be generalized, it is certain that while women remain an underrepresented group they will need moral support. This can be provided through peer counselling and through networking and mentoring programs such as those offered by women's networks, at local and international levels, through the International Network of Women Engineers and Scientists and its member associations around the world (INWES 2002). Contact with women who work as engineers in the community would help the students to understand the workplace, its difficulties, and challenges. It is also time for male professors to become role models for male students by showing a fair and respectful treatment of everyone in the class, regardless of gender, race, or sexual orientation. Finally, women students and professors should not blatantly deny that there are gender inequities in engineering. This attitude makes them part of the problem instead of part of the solution.

CONCLUSION

The model described here represents an attempt to provide students in engineering with a wide range of skills that complement their technical abilities and that meet the new demands of employers and of society. This proposal is far from being comprehensive and many other techniques can be added to the ones described in this paper. It is not necessary to add more courses to the already demanding workload of this degree, but simply to change the way in which we teach the courses that already exist. This can be done with some changes to content and by using a delivery style that is inclusive and that reaches students with different learning styles. Inviting women (and men) role models to complement our instruction with real-life and relevant examples taken from local working professionals is also helpful.

Changes to the engineering curriculum and the faculty environment will only be possible if engineering students and professors, as well as employers of engineers and professional associations make some effort to redress inequities and find original and innovative ways to include more women in all facets of engineering professional life around the world. In this rapidly changing world, success will be linked to the encouragement of diversity, and increasing the number of women engineers will help to ensure the development of a gender-balanced view upon which to develop the technological solutions of the future.

REFERENCES

Asking different questions: women and science. 1996. Directed by G. Basen and E. Buffie. 57 min. Artemis Films/National Film Board of Canada. Videocassette.

Belenky, M.F., Clinchy, B.M., Goldberger, N.R., and Tarule, J.M. 1986. *Women's ways of knowing.* New York, Basic Books.

Bouchard, M. 1991 *L'indispensable contribution des femmes en génie.* Vice-présidente, Groupe DMR, Montréal. Presentation at the Montreal Forum of the CCWE.

CCPE (Canadian Engineering Human Resources Board, Canadian Council of Professional Engineers) *Canadian Engineers for Tomorrow: Trends in Engineering Enrolment and Degrees Awarded.* Ottawa.

CCWE. 1992 *More than just numbers: Report of the Canadian Committee on Women in Engineering*, April 1992. Http://www.carleton.ca/cwse-on/webmtjnen/repomtjn.html (accessed in December 2003).

Drolet, D. and Turgeon, L. 1991 *Les filles en génie: Des différences à connaître et à respecter*. Université de Laval, Québec. Presented at the Montreal Forum of the CCWE.

Erhart, J.K. and Sandler, B.R. 1987. *Looking for more than a few good women in traditionally male fields*. Project on the Status of Education of Women. Washington, DC: Assoc of American Colleges.

ETAN Report: Report of the European technology assessment network. Available on: http://www.cordis.lu/etan/src/topic-4.htm (Accessed in December 2003).

Florman, C. 1987. *The civilized engineer*. New York: St Martin's Press.

Frize, M. 1996. *Teaching ethics and governance of the profession: Ahead or behind professional practice realities*. Proceedings of the Tenth Canadian Conference on Engineering Education, Kingston.

Gilligan, C. 1982. *In a different voice: Psychological theory and women's development*. Cambridge, MA: Harvard University Press.

Greenberg-Lake. 1991. (The analysis Group) *Shortchanging girls, Shortchanging America*. 515 Second St. NE, Washington, DC 20002.

Hacker, S.L. 1981. The culture of engineering: Woman, workplace and machine. *Women's Studies International Quarterly* 4, 341–53.

ICWES12: Proceedings of the 12th International Conference of Women Engineers and Scientists. Ottawa, July 2002. CD-ROM, can be purchased from: http://www.inwes.org (accessed in December 2003).

INWES (International Network of Women Engineers and Scientists. Available at: http://www.inwes.org (accessed December 2003)

Light, R.J. 1990. *The Harvard assessment seminars: Explorations with students and faculty about teaching, learning, and student life*. Harvard University, Cambridge, MA 02138.

NSERC Report. 1996. Increasing participation of women in science and engineering research. Report to council. Available from: Natural Science and Engineering Research Council, 350 Albert St, Ottawa, Ontario, Canada K1A 1H5 (Fax: 613-943-0742)

Pomfret, A. and Gilbert, S.N. 1991. Did they jump or were they pushed? Gendered perceptions and preferences affecting attrition among science and engineering students. (CCWE Conference), Dept. of Sociology and Anthropology, University of Guelph, Guelph, Ontario.

Robinson, G.J. and McIlree, J.S. 1991. Men, women, and the culture of engineering. *The Sociological Quarterly*, 32 (3), 403–21.

Rogers, Pat. 1988. Gender differences in mathematical ability – perceptions vs performance. Sixth International Congress on Mathematical Education, Budapest, July.

Rosser, S.V. 1990. *Female-friendly science: Applying women's studies methods and theories to attract students*. New York: Pergamon Press.

- 1997. *Re-engineering female friendly science.* New York: Teachers College Press, Columbia University, New York.
Sandler, Bernice R. 1986. *The campus climate revisited: Chilly for women faculty, administrators and graduate students.* Association of American Colleges, 1818 R St., N.W. Washington, DC.
Sorensen, K.H. 1992. Towards a feminized technology? Gendered values in the construction of technology. *Social Studies of Science* 22 (1), 5–31.
Tobias, S. 1990. *They're not dumb, they're different.* Research Corporation, 101 North Wilmot Road, Suite 250, Tucson, AZ.
Tonso, K 1997 *Violence(s) and silence(s) in engineering classrooms. Advancing women in leadership.* Available at: http://www.advancingwomen.com/awl/spring97/awlvl_2.html (accessed December 2003)
Vanderburg, W.H. 1990. Complementary studies with a global perspective. Proc. Seventh Canadian Conference on Engineering Education, 374–82, Toronto.
Vickers, M.H., Ching, H.L. and Dean, C.B. 1995. *Do science promotion programs make a difference?* Proceedings of the More than Just Numbers Conference, Faculty of Engineering University of New Brunswick, Fredericton, NB, Canada.

14

Evaluation Matters: Creating Caring 'Rules' in the Human Science Paradigm in Nursing Education

ALEXANDRA McGREGOR

There has been no dearth of studies of the professional socialization of nurses. By far the greatest amount of research has examined the degree to which nursing students and graduates from various types of nursing schools have internalized professional values and attitudes. Failures to meet standards for professional nursing practice have been attributed either to a flawed educational process or, more commonly, to the personal characteristics of recruits who are deemed not competent in their praxis. This "deficit model" approach (Acker 1983), which does not take the lived experience of failure into account, has dominated nursing discourse on education. In this model, the failing student from whom we distance ourselves becomes "the other" (McGregor 1996) and the system is allowed to continue almost sacrosanct.

As schools of nursing respond to the clarion call for nursing education to embrace notions of emancipatory education by developing nursing curricula within the caring or human science paradigm, a deeper understanding of how student successes and failures are managed and negotiated within a caring context is of central importance. From this problematic, several critical questions emerge which are explored in this chapter. First, can "caring rules" (procedural guidelines; evaluation policies and practices; ways of being with failing students) be uncovered, and what might these rules look like in contrast to traditional paradigms? Second, if caring rules can be constructed, how might these rules be enacted in practice so that students and teachers can experience teaching and learning in a more caring community? Third, when tensions about "performance" arise, how can teachers become partners who stand *with* rather than *against* vulnerable students who are struggling to become competent nurses? In highlighting what really matters in evaluation in nursing education, I pose collaborative strategies for transformative teaching and learning practices that can affirm personal and professional growth.

BEGINNING WITH MY OWN STORY:
"ATTUNED TO THE MOMENT"

Our stories of teaching speak of times when things go awry and when, through engaging and connecting in significant moments with a student, we transform and are transformed (Diekelmann 1991). I begin with a story which was a defining moment in my own teaching in nursing (Diekelmann and McGregor 2003).

I was a clinical teacher on a surgical unit where first-year diploma nursing students were caring for two or three clients in varying stages of postoperative recovery. For Paul, who already held a Bachelor of Science degree, becoming an A+ student in his theory courses was an easy accomplishment. However, his academic brilliance faded once he came to the bedside. After weeks and weeks of working closely with him, Paul's inability to transform his disorganized, marionette-like performance into a passing clinical grade had become glaring to me. With few remaining clinical days to prove his mettle, I was seeing him (and myself) as a failure. Organizing safe, competent care, even for just one client, seemed to be beyond him. We both tried to hide our frustration. Paul was more successful than I.

On one of our few remaining clinical days together, after I had decided with some degree of certainty that he was a "clinical failure," Paul and I shared a startling and illuminating few moments. We were positioned discreetly in the hallway engaged in one of our frequent hospital corridor conversations in another "teacher speak" session on Paul's performance deficits and how he should fix them. Seemingly out of nowhere, Sister Katherine, an acquaintance from my previous years in management at that same hospital, rushed up to me and effectively gave me a jolt that to this day I have been unable to forget. "Oh, I am so glad to see you," she began loudly, and before I could even interject to say hello or gently indicate that I was in the midst of a student dialogue, she proceeded with a stream of words that literally sent a shock through my being. "I had the same surgery you did ... They took my breast too. Seeing people like you helps me believe I will be alright too." Almost surreally, with a smile and a quick wave, she disappeared down the hall.

Comprehension of the full context of this public moment requires exploration of what this intrusive revelation meant to me. I am, and was, a fairly private person. While generally warm and open in my interactions, I had chosen to deal with my cancer and subsequent mastectomy, more than 10 years earlier, with the support of close friends and family privately. I had met my struggle head on and moved through it, or so I thought. So now, here I was, standing in the hallway with a student who in my mind "just couldn't cut it" and feeling exquisitely exposed and horrified that my personal world had been cracked open with these few careless words. Stunned momentarily into silence, I looked up at Paul who was no longer only a student, but also a witness. Without a moment's hesitation, with gentleness, and with no hint that anything was amiss, he brought me back to the present. "Alix," he started calmly, "you were suggesting ways to plan my care for tomorrow." Paul hadn't *missed* the meaning of the brief but personally devastating exchange. Instead, Paul had "got it," and he'd gotten it so completely that he responded not only as a person and as a student, but even more significantly, as a nurse.

Had I not been attuned to the moment, I might have missed that critical insight. Intuitively, he'd managed to nurse me back to a place where I felt safe once again. And from that place of safety, from that caring moment, I was able to look at Paul with eyes that allowed for a greater depth of experience. With the memory of my own vulnerability giving me the gift of context, I was able to reach across the distance between myself as 'teacher' and Paul as 'student.' In crossing that chasm, I was able to see Paul not just differently, but with more clarity. I was able to see, because I had been able to experience in that brief moment that he was a nurse rather than a failing nursing student. Consciously and deliberately, I "let him be" for the remaining clinical days to nurse his patients in his own way. From that moment in the hallway, he managed just fine without my oppressive scrutiny and guiding "rules" and he earned a passing clinical grade.

In the early 1980s, "crossing the chasm" between students and teachers was not unlike swimming upstream against a traditional system which guided "knowing" only through *power-over* (that is, mechanistically, with an emphasis on measurement of objectives), as students' clinical grades were determined by the faculty. However, from the beginning of the curriculum revolution in 1990 which called for caring to be the core value in nursing curricula (Beck 2001), there has been a paradigm shift from a behaviourist (traditional, positivist, or natural science) approach to a human science (non-traditional or naturalist) approach.

Paterson and Zderad (1988) developed a humanistic theory of nursing, upon which other nursing theorists have drawn and, increasingly, nursing is being referred to as a human science (Mitchell and Cody 1992). The central argument propelling adoption of the human science paradigm (in academe and practice) is that a passive, 'banking' socialization process (Friere 1970) oppresses both students and teachers, serves to maintain the status quo, and renders nurses only capable of fitting into a "wounded" health care system rather than becoming agents of social change. By the end of the decade, it had become apparent that traditional methods of evaluation not only failed to support students and faculty through the actual evaluative process, but they also failed, essentially, to support the unique values at the very core of nursing. This dissonance, palpable for both students and faculty, needed to be addressed.

PRIMACY OF THE STUDENT-TEACHER RELATIONSHIP

With the development of nursing curricula that embrace the caring or human science paradigm (Bevis and Watson 2000; Hills and Lindsey 1994), oppressive relations of power between students and teachers have come under scrutiny. In recognizing that nursing teachers fulfill multiple and seemingly conflicting roles (caring mentor, teacher, evaluator, and gatekeeper of the profession), doors are opening to evaluate learning differently. While the curriculum revolution literature proposes emancipatory ways of being which can transform power relations between students and teachers (Romyn 2001), this discourse fails to take fully into account the shifting vulnerability in these relationships. The heart of this vulnerability, particularly for practice disciplines, can be found when a student's academic or clinical performance falls below program expectations. Moral conflicts arise as faculty, who embrace caring as their mantra, find themselves coming up against situations in which this ideal, if not flawed, is at least limited in its applicability.

Living a caring-based program where student evaluation is grounded in "a process of valuing, prizing, and growing," Schoenhofer and Coffman (1994, 152) posit that, when students fail, it is "how students and teachers are with each other and how they proceed in [their] relationship" that changes. Their claim is in sharp contrast with the more traditional procedural "educational ideals" which inform clinical evaluation approaches and guide decision-making abut unsafe students (Rittman and Osburn 1995; Scanlon et al. 2001) and legal issues related to clinical failures (Orchard 1994; Smith et al. 2001). Obsession with improving objectivity in clinical evaluation has generated a plethora of mechanistic, competency-based evaluation tools, none of which have served to resolve the objective-subjective debate. Nor do they take into account multiple ways of knowing and being. Distinctly absent is the notion of caring in this discourse, despite the fact that the caring curriculum is widely accepted throughout programs of nursing education. So, here again, the dissonance continues, although perhaps in a more subtle, and more pervasive manner.

The terms *formative* (educative, growth-promoting) and *summative* (gate-keeping, grade awarding) convey the dualistic nature of clinical evaluation and highlight a central tension in the student-teacher relationship. Clinical grade determinations require synthesis of data collected, forcing the teacher to step out of the subjective mode towards the end of the term. As Mahara comments, "the artificial separation of teaching-evaluating and formative-summative (evaluation) creates a dichotomy that serves to maintain distance between teacher and student and perpetuates a belief that students are unable to be 'knowers' of quality practice and their own learning" (1998, 1341).

Paradoxically, the human science paradigm, which celebrates the primacy of the teacher-student relationship, which promotes egalitarian or shared responsibility for learning (Tanner 1990), and which positions the teacher as learner with the students, can break down when students are borderline or failing. Institutional procedural arrangements cast the caring, egalitarian teacher, who within this human science paradigm has begun to view himself or herself almost romantically as teacher-learner, into the more rigid, less idealistic role of teacher-evaluator during the summative or grading phase of the evaluation process from which the student is removed. The assumption is that, when final decrees are determined, the power and expertise lie solely with the teacher. As Duke's (1996) research identified, faculty, particularly in the case of sessional or inexperienced nursing faculty, had difficulty failing nursing students with whom they had developed a "caring relationship." Negligent behaviour was accepted when viewed within the context of the student's personal world. Nursing faculty face moral conflicts as they balance the well-being of students, clients, and the nursing profession. In making pass-fail decisions, faculty understand at their core that "we are mucking around with people's lives here" (McGregor 1996). Herein lies one of the great points of tension for nurse educators. Passing or failing a student may literally translate into a life or death decision when the responsibility students will take on upon graduation is considered. While no doubt this added responsibility inherent to the evaluative component of nursing education has historically supported the traditional, prescriptive methods previously described, there are also very significant detractors to this approach.

RUINING THE PERFECTLY "GOOD" STUDENT: PERSPECTIVES ON ACADEMIC AND CLINICAL EVALUATION

And so the question arises, does the caring pedagogical framework make a difference in the lived experiences of students and faculty when negotiating academic and clinical success or failure? Returning to my experience with Paul when I became "attuned to the moment," the central meaning now for me is that, without that chance encounter and my openness to its significance, I would have failed Paul because he did not fit my commonsense view of the 'good' student. Caught as I was in the systemic pressure to provide students with solid medical-surgical experiences stuffed with skill-based nursing interventions, his essential core was invisible to me. Coming to view Paul as a nurse enabled me to halt from my distancing, "walling off" (Malone 2000) stance and to bring forth a "face-to-face connectedness " (Berman 1994). Though I was unable to name my pedagogy as "caring" at the time, I now recognize that Aoki (1992, as cited in Berman 1994) captures the heart of my teaching within the human science paradigm: "[The teacher] sees pedagogic teaching not so much as leading that asks followers to follow the leader assuming that the leader always knows the way. Rather, she[/he] sees it as a *responding responsibility* [my emphasis] to students. Such a leading entails at times *letting go* that allows a *letting be* in students' own becoming" (Berman 1994, 173). This "letting go" approach represents a marked shift from the traditional, "power-over" oppressive approach in which the professor assumes that "it's only right if you do it my way" (McGregor 1996). While most students survive their socialization journey – some by becoming authentic, some by doing it "the teacher's way for now," and some by letting harsh realities "roll off like water off a duck's back" – we are at risk of inadvertently preventing perfectly 'good' students from reaching their optimal potential.

Over my years of teaching nursing, and more recently in a caring-based curriculum, students have revealed conversations with their teachers in clinical and classroom situations from which they have emerged feeling "less than" rather than "more than." These interactions, which students experience as uncaring and unhelpful, have the potential to destroy even the most resilient student. Always saddened by these stories of disjunction which cast our caring ideal into the realm of rhetoric, my own sense of vulnerability surfaces strongly as I, in striving and longing for this almost Utopian ideal, find myself falling short. Though my own story highlights a special time when, in Dieklemann's words (1991), "I got it right," I wonder about those times when, knowingly or unknowingly, "I got it wrong."

SHIFTING VULNERABILITY

In my research, what I would call an exquisite vulnerability for nursing students and faculty was revealed as they attempted to understand and manage failure. From Olesen's explorations of individuals' mundane experiences (of minor ailments) juxtaposed with extreme external events (a California earthquake), the concept of a "renewed vulnerable self" emerged (1992, 205). Continual personal transformations occurred "where [the] self reflects upon self." In nursing, where experiential learnings are anchored in external situational contexts from which personal reflection and introspection follows, perspective transformations do occur as nursing students take on successful or failing identities. Both "successes" and "failures" can make one feel vulnerable. Building on this notion, "we could explore how vulnerability emerges, how it feels in terms of emotions, what it means, how it becomes part of biography, and how reflexive processes lead to a vulnerable self" (Olesen 1992, 217). From the students' perspectives, when they receive muddied or clear messages from their peers, nurses, or faculty that their "attitude" is unbefitting a nurse, that their theory base is weak or integrated inadequately, that their demeanour is not suited for nursing, or that their performance as a nursing student is a failure, a vulnerable self emerges as the meaning of these experiences and their consequences are explored.

From a faculty perspective, what kinds of events trigger a "reflexive process" that leads to the emergence of a vulnerable self? Ironically, on the one hand, nursing students commonly mistrust faculty relationships and their clinical evaluations (McGregor 1996). On the other hand, faculty fear that nursing students who pose a risk to future clients may be graduated or that their evaluations may not be upheld by their colleagues. They fear personal attacks when assigning a failing grade and being labelled "a failing teacher" (Carpenito 1983). They fear that their appraisals of students may be flawed or unfair, or that their judgements may come under the scrutiny of a student appeal process. The risks of failing a nursing student, particularly one for whom the decision to fail is not patently obvious, can thus create a deep sense of vulnerability for faculty and the institution. Delivering a failing evaluation and discontinuing a nursing student are the most difficult institutional tasks to perform, particularly when accompanied by marked resistance from the student. As one former colleague commented: "Maybe it's a reflection on themselves [students] and partly you feel like a failure because you are so concerned with that individual. It's not just that they have failed nursing. You have affected someone's life. This is a big deal. This isn't something you take lightly" (McGregor 1996).

Despite successful moments in which we as faculty take momentary pride in our own movement towards ideological and pedagogical congruence in our practice, it can take only one troublesome failing event to evoke a personal sense of failure in ourselves. Embedded in the question of agency and our role in the structuring of failure, this emergence of our vulnerable self echoes the students' sense of "doing good but feeling bad" (McGregor 1996).

Despite caring (or non-caring) procedural guidelines outlining the standards for evaluation in nursing practice, nursing faculty express a great deal of anguish in their decision-making about what to "jump on" and what to let go. Perhaps it is as Orchard (1994) has suggested, that the rights of students have superseded the rights of nursing faculty in evaluating student success and failure. Resolving realistic fears of wrongfully triggering a "spoiled identity" (Goffman 1963), "letting go" when on reflection they needed "to jump," negotiating student-faculty relationships within the context of shifting vulnerabilities, pressing for student success despite multiple systemic barriers, and responding to the institution's proclivity for extensive documentation in the face of failure all serve to challenge the faculty member's sense of personal control. As the question "whose failure is it?" surfaces for some faculty, the fear that it is theirs may roll back in waves of self-doubt.

In my research, as I approached the question "whose failure is it?" a "vulnerable self" emerged from the voices of the faculty. Building a "case" of failure brings forth a mixture of maternal-like caring burdened by feelings of guilt and self-doubt, as this faculty member's reflections (which she left hanging) suggests: "If I can sit back and say I've done everything possible to try and help a student meet those objectives, then

I shouldn't be thinking, feeling – [lengthy pause]. I mean you do feel badly because you like them. And we don't want to see anybody not succeed ... so it's not, it's not easy even though we know it is the best thing for them. You know, it's like telling a child to take some medicine that doesn't taste very good" (McGregor 1996).

Failure in nursing is painful for all of the players, bringing forth feelings of being exposed and unprotected. Denying ownership of the failure can bring comfort for a time, and yet, the question "whose failure is it?" comes back again and again.

DANCE OF THE DOMINATOR: "GETTING THROUGH" TO THE STUDENT

> knowledge of the oppressor
> this is oppressor's language
> yet I need it to talk to you.
> (Rich 1975, 149)

Now a central theme in the dominant discourse in nursing education, the human science or caring paradigm is purported to foster student-faculty relationships in which former power imbalances are rectified and in which what I call "caring rules" emerge through a process of negotiation with the students. The language of learning, becoming, being in equity and equality takes on an almost surreal quality, as Boykin's metaphor of the "dance of nursing" portrays: "Persons sharing in the dance of nursing [are in] a circle and not a hierarchy, all persons at the same level. Therefore, there is an opportunity to truly know each other as [a] caring person. Each dancer is recognized, valued, and celebrated for the gifts he or she brings. No one role is more important than the other" (1994, 17).

The notion of "getting through to the student" is clearly absent, and "oppressor's language" does not belong in the dance. And yet, when nursing students write failing papers or falter in their praxis, what happens to the equal footing? How can faculty sustain the circle of care in the face of failure? According to Diekelmann (2001), when students do not meet academic or clinical performance standards, then "you [faculty] return to conventional pedagogy" with a velvet hammer so that the students understand that they are out of step with what constitutes success in the eyes of the nursing profession. This counsel mirrors Smith's (1987) "ruling relations" construct. Concretized expectations, standards, text-based rules supersede the notion of caring, leaving the student feeling double crossed in the rhetoric of celebrating the collective "gifts" we bring to the encounter. It seems that ethereal caring is only for the successful while conventional pedagogy is reserved for students on the fringe.

While faculty would prefer that students enter with a "clean slate ... otherwise they [students] mess things up for us" (Diekelmann 2001), the reality is that our students are not always going to model our ideal notions of the good student. And because the nature of nursing calls for indisputable and lofty standards, nursing faculty are compelled to "lower the boom" when bearing witness to a potentially unsafe or failing performance. In the face of this kind of legitimized power, faculty cannot turn away from their responsibility when a student is not "fitting in." Through one student's eyes, both faculty distancing and disdain can be seen: "There is a certain amount of disdain [among faculty] for anybody on the edge around here. There's a big group – you want to be standing in the middle of it. Catch a lot less crap. It's the people on the edges that we've seen going down over and over again. It's always the people that didn't quite fit in from the beginning that aren't making it" (McGregor 1996).

The image of students "going down over and over again" is in sharp contrast with a caring-based nursing program in which "evaluation processes are recognized as opportunities to grow, rather than being addressed as token compliance or threats to the status quo" (Schoenhofer and Coffman 1994, 129). We rein in students from the edges, shaping their performance with 'success plans' – sugar-coated caring rules which hide real threats of failure to meet course or program outcomes. It seems to me that failing within a caring curriculum sets faculty and students up for a greater fall. The implication is that neither the faculty nor the student simply could care enough, evoking so strongly what I have referred to as that exquisite sense of vulnerability.

BEARING WITNESS TO FAILURE: STANDING WITH THE STUDENT

I can recall conversations with former colleagues around the lunch table in which they expressed open apprehension (sometimes with humour) about having to deliver a failing grade to students who were not making it. Referring to imminent evaluation meetings as "turtleneck days," one colleague, who modelled caring for students in exemplary ways, laughingly explained that she needed her turtleneck sweater "to hide the blotches ... I can't let them see how nervous I am in giving them the bad news." Rather than uncovering the real story of shared vulnerability, faculty find themselves "walling off" their emotions as they struggle with a host of uneasy feelings. Touching briefly on her unease, this colleague highlights some of the layers of faculty fears about evaluation: "It's not easy and I mean you're sitting there eye to eye telling them 'I'm sorry but, you know' [pause], often faculty who do it [fail students] will say: 'All the students hate to get me' and 'I've got a real reputation around here.' Unfortunately, after awhile, people stop doing it [failing students] which is frightening too. I myself have gotten students who've gotten through a previous semester and are supposed to be at certain level and are *way, way back there* and you sort of say to yourself: 'Hey, wait a minute. How did they get through?'" (McGregor 1996)

Caring-based programs call for "authentic presencing" in relationships with students so that in failure, students know that faculty are "standing with" them rather than against them. When facing failing events in students lives, "faculty and students are challenged [to find] realistic ways of standing with, of living caring values such as honesty, humility, knowing, presence, and connectedness in the evaluation process" (Schoenhofer and Coffman 1994, 129). In my relationship with Paul, I found a way of "letting go"; yet, had I not been able to do this, I believe now that he would have remained "faceless" as I delivered a failing clinical grade. Certainly, I would have been seen as standing against him. In this situation, the path to consensual meaning making, that is, mutually coming to understand that he could not pass, would have been blocked by my predetermined interpretation of the semester's events. In an ideal world, the student and teacher would come to the realization that their "readiness to move on" is not there yet. Recalling a student's experience with failure, Diekelmann (2001) shared a mother's comments to her daughter: "You haven't failed nursing ... you just haven't passed yet". And so the question becomes "what are the possibilities for honouring our shared vulnerability and being truly present for and with one another?" Somehow, we have to find ways to live with failure, offering respect and kindness to those who are lost "on the edge," and offering opportunities "for keeping open a future of new possibilities" (Diekelmann and McGregor 2003).

CONCLUSIONS

Caring rules is clearly an oxymoron. For those of us who are co-creating and living with an emancipatory curriculum, multiple challenges emerge along the way as we strive to meet our obligation to uphold exalted academic and professional standards. When students are exemplary and able to "prove" their remarkable competence and success, all is well and we rejoice fully in those accomplishments. Yet, as Maraha (1998) points out, questions about how to separate teaching from evaluation in an emancipatory framework seldom surface with successful students. When Schoenhofer and Coffman (1994) speak to the challenges of being in relationships with *unsuccessful* students and allude to how these ways of being are "different" from traditional paradigms, these ways are framed in caring rhetoric. Faculty are admonished to be authentic and fully present in their relationships with failing students. Compelling as the arguments are, they do back away from offering real or practical ways of being which would support faculty and students who struggle with performance issues in every semester. Clearly, research that can illuminate failing experiences and that can guide best practices in evaluating students is needed.

According to Beck (2001), authentic presencing opens the way for caring to unfold. As nursing teachers, "being called to care" (Lashley et al. 1994) for failing students is not easy. While we may want to care for students as we witness their failure, we may not be alone in the need to "wall off," to escape the pain and public humiliation of failing. We can leave the door open, but the "letting go" also means accepting students' choices, even if this means that our caring is rejected or they become marginalized through the very choices they make.

Those who embrace and attempt to live out the ideals of an emancipatory curriculum face many challenges. Learning to live in an imperfect world with imperfect systemic structures is not easy, particularly when students' personal and professional growth plays out in ways other than what we might wish. Can we truly let go, allowing a "letting be in a students' own becoming" when it might not be our way?

REFERENCES

Acker, S. 1983. Women and teaching: A semi-detached sociology of a semi-profession. In *Gender and education*, eds. L. Barton and S. Walker, 123–39. Lewes, England: Falmer Press.

Beck, C. 2001. Caring within nursing education: A metasynthesis. *Journal of Nursing Education* 40(3), 101–09.

Berman, L. 1994. What does it mean to be called to care? In *Being called to care*, eds.

Lashley, M.T., E. Slunt, L. Berman and F. Hultgren, 5–16. New York: State University of New York Press.

Bevis, E. and Watson, J. 2000. *Toward a caring curriculum: A new pedagogy for nursing*. Sudbury, MA: Jones and Bartlett.

Boykin, A. (Ed.). 1994. *Living a caring-based curriculum*. New York: National League for Nursing, 11–25.

Carpenito, L. 1983. The failing or unsatisfactory student. *Nurse Educator* 8(4), 32–3.

Diekelmann, N. 2001. Personal notes from collaborative BScN nursing faculty retreat with Dr N. Diekelmann, York University School of Nursing, Toronto, November 14–15.

– 1991. The emancipatory power of the narrative. In *Curriculum revolution: Community building and activism*. (National League of Nursing Publication No. 15-3298), 41–62. New York: National League of Nursing.

Dieklemann, N. and McGregor, A. 2003. Students who fail clinical courses: Keeping open a future of new possibilities. *Journal of Nursing Education* 42(10), 433–36.

Duke, M. 1996. Clinical evaluation – difficulties experienced by sessional clinical teachers: A qualitative study. *Journal of Advanced Nursing Science* 23, 408–14.

Freire, P. 1968/1970. *Pedagogy of the oppressed* (M. Ramos, trans.). New York: Continuum (Original work published in 1968).

Goffman, E. 1963. *Stigma: Notes on the management of a spoiled identity*. New York: Doubleday.

Hacker, S. 1990. Doing it the hard way. In *Investigation of gender and technology*, eds. D. Smith and S.M. Turner. Boston: Unwin Hyman.

Hills, M. and Lindsey, E. 1994. Health promotion: A viable curriculum framework. *Nursing Outlook* 43, 158–62.

Lashley, M.E. 1994. Vulnerability: The call to woundedness. In *Being called to care*, eds. A. Lashley, M. Teal, E. Slunt, L. Berman and F. Hultgren, 41–52. New York: State University of New York Press.

Mahara, M.S. 1998. A perspective on clinical evaluation in nursing education. *Journal of Advanced Nursing* 28 (6), 1339–46.

Malone, R. 2000. Dimensions of vulnerability in emergency nurses' narratives. *Advances in Nursing Science* 23 (1): 1–11.

McGregor, A. 1996. *The professional socialization of nursing students: Failure as a social construction*. PhD dissertation, University of Toronto, Toronto, ON.

Mitchell, G. and Cody, W. 1992. Nursing knowledge and human science: Ontological and epistemological considerations. *Nursing Science Quarterly* 5(2), 54–61.

Olesen, V. (1992). Extraordinary events and mundane ailments: The contextual dialectics of the embodied self. In *Investigating subjectivity*, eds. C. Ellis and M. Flaherty, 205–20. Newbury Park: Sage.

Orchard, C. 1994. The nurse educator and the nursing student: A review of the issue of clinical evaluation procedures. *Journal of Nursing Education* 33(6), 245–251.

Paterson, J. and Zderad, L. 1988. *Humanistic nursing*. New York: National League of Nursing (Originally published, 1976, Wiley).

Rich, A. 1975. The burning of paper instead of children. In *Poems: Selected and new, 1950–1974*. New York: W.W. Norton.

Rittman, M.R. and Osburn, J. 1995. An interpretive analysis of precepting an unsafe student. *Journal of Nursing Education* 34: 217–221.

Romyn, D. 2001. Disavowal of the behaviourist paradigm in nursing education: What makes it so difficult to unset. *Advances in Nursing Science* 1: 1–10.

Scanlon, J., Care, W. and Gessler, S. 2001. Dealing with the unsafe student in clinical practice. *Nurse Educator* 26(1), 23–7.

Schoenhofer, S. and Coffman, S. 1994. Prizing, valuing, and growing in a caring-based program. In *Living a caring-based program*, ed. A. Boykin, 127–65. New York: National League for Nursing.

Smith, D.E. 1987. *The everyday world as problematic: A feminist sociology*. Boston: Northeastern University Press.

Smith, M., McKoy, Y., and Richardson, J. 2001. Legal issues related to dismissing students for clinical deficiencies. *Nurse Educator* 26(1), 33–8.

Tanner, C. 1990. Reflections on the curriculum revolution. *Journal of Nursing Education* 29(7), 295–99.

15

Post-colonial Remedies for Protecting Indigenous Knowledge and Heritage

MARIE BATTISTE

GENERATING POST-COLONIAL REMEDIES

Many universities and other educational institutions state in their mission statements or in their institutional goals that the education of Indigenous peoples is a priority; however, to date these priorities have been neo-colonial. In other words, Indigenous people can access what is available, but they cannot change the existing knowledge base. In order for educational institutions to live up to their alleged priorities, they must introduce post-colonial frameworks into their curricula. To do this, educators must confront the colonial history of education, the Eurocentric content of current curricula, and the attitudes of superiority that continue to demean the role of Indigenous knowledge and people in education. This sense of superiority is especially evident in the public perception that Indigenous ownership, custodianship, and management of their territory and resources, indeed of their knowledge and education, is conflict-prone and incompatible with contemporary standards of education and society. Post-colonial education offers some hope to Indigenous people that a solution to cognitive imperialism will soon be found.

The decolonization of Eurocentric thought is already underway in the works of many scholars; however, the experiences of Indigenous people engage decolonization in a distinct manner. Maori educator and scholar

Linda Tuhiwai (too-ee-way) Smith, one of the leading theorists of decolonization in New Zealand, clarifies the nature of the task when she writes: "Decolonization is about centring our concerns and world views and then coming to know and understand theory and research from our own perspectives and for our own purposes" (1999, 39). The interrelated strands of scholarship and experience intersect to weave solutions not only to decolonize education, but also to sustain the Indigenous renaissance and to empower intercultural diplomacy. Renaissance and empowerment are two essential features of the post-colonial movement. Post-colonialism is not about rejecting all theory or research of Western knowledge. It is about creating a new space where Indigenous peoples' knowledge, identity, and future is calculated into the global and contemporary equation.

Decolonization has both a negative movement and a positive, proactive one. The negative movement has been clearly articulated. As Dr Daes noted at the UNESCO Conference on Education in July 1999: "Displacing systemic discrimination against Indigenous peoples created and legitimized by the cognitive frameworks of imperialism and colonialism remains the single most crucial cultural challenge facing humanity. Meeting this responsibility is not just a problem for the colonized and the oppressed, but rather the defining challenge for all peoples. It is the path to a shared and sustainable future for all peoples" (1999, 1).

It is becoming clear that attempting to decolonize education and actively resisting colonial paradigms is a complex and daunting task. Educators must reject colonial curricula that offer students a fragmented and distorted picture of Indigenous peoples, and offer students a critical perspective of the historical context that created that fragmentation. In order to effect change, educators must help students to understand Eurocentric assumptions of superiority within the context of history and to recognize the continued dominance of these assumptions in all forms of contemporary knowledge.

Those who are researching Indigenous knowledge must understand both the historical development of Eurocentric thought and Indigenous contexts. A body of knowledge differs when viewed from different perspectives. Not only will interpretations or validations of Indigenous knowledge depend on the researcher's attitudes, capabilities, and experiences, but they will also be based on the researcher's understanding of Indigenous consciousness, language, and order.

Depending on the Eurocentric reductionistic analysis used, Indigenous knowledge may be segmented or partial, utilitarian, or non-utilitarian, or both. Indigenous knowledge needs to be interpreted based on form and manifestation, as the Indigenous people themselves understand them.

According to historian Lise Noël (1993), domination and oppression are grounded in intolerance. The fact that the modern intolerance is Eurocentric consciousness itself has profound implications for the liberation of Indigenous knowledge. In terms of getting educated, finding knowledge, doing research, and constructing research ethics, where are Indigenous people to find experts who can rise above the value contamination of their consciousness? Where are these people being trained? By what faculty are they being taught? Because of the pervasiveness of Eurocentric knowledge, Indigenous peoples do not have at their disposal any valid ways to search for truth.

Every university discipline (and its various discourses) has a political and institutional stake in Eurocentric diffusionism and knowledge. Every university has been structured to see the world through the lens of Eurocentrism, which opposes Indigenous knowledge. The faculties of contemporary universities encourage their students to be the gatekeepers of Eurocentric disciplinary knowledge in the name of universal truth. Yet, Eurocentric knowledge is no more than a Western philosophy invested in history and identity to serve a particular powerful interest. When it approaches Indigenous issues or peoples, its research methodology is contaminated with multiple forms of cognitive imperialism.

All peoples have knowledge, but the group that controls the meanings and diffusion of knowledge exercises power and privilege over other groups. Those with control then link the diffusion of knowledge with economics, ensuring that some kinds of knowledge are rewarded and other kinds of knowledge are not. Public schools are not politically neutral sites. Using sanctions, grants, and awards, governments tie knowledge to their own interests, using vague notions of 'standards' and 'public good' to control what counts as knowledge, how this knowledge is diffused, and who benefits from it. In all cases, those holding colonial power control knowledge, and Indigenous knowledge is excluded from curricula as it is regarded as too local or too particularistic, or inferior to universal knowledge. It is only recently that ethno-botanists and pharmaceutical companies have begun to see the potential and value in buying or appropriating Indigenous knowledge, patenting the results, and selling them back to Indigenous peoples in the form of medicines and herbal remedies. This commodification of Indigenous knowledge occurs where knowledge, power, and economics intersect, which takes me to the theme of political voices and visions of Indigenous peoples.

In 1989, a report from the United Nations Seminar on the Effects of Racism and Racial Discrimination on the Social and Economic Relations between Indigenous Peoples and States concluded that Indigenous peoples are being subjected to a new form of global racism (UN 1989). This modern form of discrimination comes under the guise of state theories of cultural, rather than biological superiority, and it results in rejection of the legitimacy or viability of Indigenous peoples' own values and institutions. Often, theories of both cultural and biological superiority are involved and interconnected in responses to Indigenous peoples, and cultural racism is alive and well in the areas of intellectual and cultural property rights.

Discrimination is defined both internationally and nationally as any unfair treatment of, or denial of normal privileges to, persons because of their race, age, sex, nationality, or religion. When discrimination is effected through the machinery of the state, it can have devastating impacts, ranging from deep psychological scars to racial and cultural genocide. For victims of discrimination, it matters little whether the damage is inflicted by invidious state action or by the less obvious application of facile neutral rules. The impact is the same. Not only should nations not practise discrimination themselves, they must also identify ways in which they will protect their citizens from discrimination on both local and national levels.

CREATING THE ABORIGINAL RENAISSANCE

A post-colonial framework cannot be constructed unless Indigenous people renew and reconstruct the principles underlying their own world views, environments, languages, and forms of communication, and re-examine how all these elements combine to construct their humanity. I have dealt with this process of reclamation in *Reclaiming Indigenous Voice and Vision* (Battiste 2000). Newly empowered Indigenous people and their non-Indigenous allies are providing critical frameworks for addressing these issues while acknowledging excellence through the proper valuing and respectful circulation of Indigenous knowledge across and beyond Eurocentric disciplines. Indigenous people are seeking to heal themselves, to reshape their contexts, and to effect reforms based on a complex arrangement of conscientization, resistance, and transformative action.

Through collaborative work with scholars in Canada and beyond to Australia, New Zealand, and the United States, Indigenous scholars and leaders are illustrating the strength of the post-colonial movement by constructing the multidisciplinary foundations essential to remedying the acknowledged failure of the current system.

Indigenous peoples' struggles cannot be reduced to singular solutions in singular locations. They need to be carried out in multiple sites using multiple strategies. Among these important sites and strategies is the work being done by Indigenous peoples at the United Nations, where lobbying bodies have raised awareness of the plight of many oppressed peoples around the world. The United Nations has produced a number of declarations and covenants embracing or urging the adoption of standards to protect women, children, and cultural minorities. Indigenous peoples remain the only people in the world without a declaration or covenant for their protection. The United Nations has provided a forum for acknowledging the concerns of Indigenous peoples, but member states have so far refused to endorse any official covenant to protect them, although the year 2004 marked the end of the Decade of Indigenous Peoples.

Many Indigenous peoples around the world continue to suffer from genocide and continue to have their lives destroyed by colonization. Efforts at the international level are intended to initiate dialogue, advance a post-colonial discourse, and work actively for the transformation of colonial thought. Indigenous peoples must work to reveal inconsistencies, to challenge assumptions, and to expose ills. They must search within themselves and their Indigenous heritage for principles that will guide their children's future in a dignified life.

In Canada, educational institutions have a pivotal responsibility in transforming relations between Aboriginal peoples and Canadian society. The Royal Commission on Aboriginal Peoples (RCAP) asserted that all institutions should consider respect for Aboriginal knowledge and heritage to be a core responsibility rather than a special project to be undertaken after other obligations have been met (RCAP, 1996, 3:515).

Multidisciplinary work is beginning to offer transcultural coalitions across education, the humanities, the social sciences, and law. Physical scientists are coming on-board as issues of biodiversity and globalization put institutions under pressure to seek innovative and holistic solutions to ecological and global problems.

It is gratifying to see the bridges being built by non-Indigenous scholars; however, respect for Indigenous knowledge must begin with Indigenous people themselves. It is Indigenous people who must provide the standards and protections that accompany the centring of Indigenous knowledge. Many of us have taken up this challenge in many forms and forums. Some notable contributions include another recent publication I co-authored with J. Youngblood Henderson, *Protecting Indigenous Knowledge and Heritage: A Global Challenge* (Battiste and Henderson 2000b) and the works of Maori scholars and educators Graham Hingangaroa Smith (2000) and Linda Tuhiwai Smith (2000) of New Zealand.

The United Nations designated 2001, the cusp of the new millennium, to be the International Year of Dialogue with Civilizations. It was a time for us to rethink the assumed values upon which contemporary society has built its knowledge base and institutions, the mythical portraits that have been assumed from this biased knowledge base, and the neglected diversity of our nations. The knowledge bases of Indigenous peoples, as well as other neglected knowledge bases, can be sources of inspiration, creativity, and opportunity. They can also contribute to humanity, equality, solidarity, tolerance, and respect. The United Nations is providing Canadian educational institutions with an exemplary opportunity to confront the ethics, methodologies, and lessons of Indigenous knowledge and heritage. As the RCAP made clear, the dialogue needed to engage the suppressed knowledge of Indigenous civilizations has been delayed for far too long. The significance and justification for respectful dialogue as a basis for arriving at a decolonized educational agenda cannot be over-emphasized. This dialogue should take us beyond the processes of cross-cultural awareness, inclusion, and bridging programs to a new perspective that supports Indigenous knowledge, communities, languages, and self-determination in a new decolonized way.

Despite the painful experiences of Aboriginal peoples in Canada over the last century or more, they still see education as the hope for their future, and they are determined to see education fulfill its promise (RCAP 1996, 3:433–4). Indigenous peoples around the world have made this clear in many voices and at many forums. Foremost among these are the reports of the RCAP, the dedicated efforts of the United Nations Working Group on Indigenous Populations, and the work of Indigenous researchers and post-colonial scholars and leaders. Institutions can no longer excuse their inactivity by saying that they don't know what Indigenous peoples want (Havemann 1999).

We must develop post-colonial strategies for both Indigenous peoples and non-Indigenous educators. We must seek appropriate protocols and respectful methodologies that will help us to enter into a dialogue and sustain respectful relations. To achieve decolonization, Canada, its people, and its institutions will need to learn to view Aboriginal peoples not as disadvantaged racial minorities, but as distinct, historical, and socio-political communities with collective rights (Chartrand 1999).

Sustaining cultural diversity is a global task. Decolonized methodologies and solutions are sorely needed to restore ecological and linguistic diversity, and respect for Indigenous peoples, their knowledge, and the continuing practices that ensure the protection of these culturally diverse societies. The challenge for Indigenous peoples is to restore their own spirits and bring back their health, their dignity, their visions, and their voices. They also need to develop strategies for revitalizing their languages, their knowledge, and their heritage in an ecologically safe and flourishing environment. This cannot happen unless nations take responsibility for the impact their destructive policies have had on their peoples and their ecologies. Nations must respect and draw upon the broad principles for decolonization that have been outlined in the International Labour Organization's Convention 169, the Indigenous peoples' polycultural *Declaration of Rights of Indigenous Peoples*, and the proposed *Guidelines for the Protection of Indigenous Heritages* (Wiessner and Battiste 2000), which outlined to the United Nations Human Rights Commission standards for establishing a fair and minimum definition of globalization.

Too many recommendations for globalization concentrate on its impact on the elite and bourgeois in the areas of economy, politics, and communication. Indigenous peoples should articulate standards for globalization that will have a positive impact on the lives of the marginalized. We might consider a litmus test for globalization that can be considered from a number of different perspectives. This litmus test should consider the impact of globalization on Indigenous languages and thought. The global village cannot survive the linguistic devastation of the last few centuries. This litmus test should also consider how globalization can create a just and mutually enriching intercultural space for respectful communication and exchange that will enhance human dignity and prosperity. Globalization has to be an inclusive complexity where none can be excluded.

The Canadian Constitution articulates the principle of maintaining respect for Aboriginal rights and treaties. Canada has a responsibility to live up to its reputation as a compassionate and innovative nation on the way to becoming a truly just society. We can only arrive at this truly just society by recognizing our dependencies on Aboriginal knowledge, values, and visions, and by renewing our investment in holistic and sustainable ways of thinking, communicating, and acting together.

REFERENCES

Battiste, M. and S. Wiessner. 2000. The 2000 revision of the United Nations draft principles and guidelines on the protection of the heritage of Indigenous people. *St. Thomas Law Review 13* (1): 383–414.

Battiste, M. and J.Y. Henderson. 2000a. *Reclaiming Indigenous voice and visions.* Vancouver: University of British Columbia Press.

– 2000b. *Protecting Indigenous knowledge and heritage: A global challenge.* Saskatoon, SK: Purich Publishing.

Chartrand, P. 1999. Aboriginal peoples in Canada: Aspirations for distributive justice as distinct peoples. In *Indigenous peoples rights in Australia, Canada, and New Zealand*, ed. P. Havemann, 88–107. Auckland, NZ: Oxford University Press.

Daes, E. 1999 (July). Cultural challenges in the decade of Indigenous peoples. Unpublished paper presented at the UNESCO Conference on Education. Paris, France.

Havemann, P. (Ed.) 1999. *Indigenous peoples rights in Australia, Canada, and New Zealand.* Auckland, NZ: Oxford University Press.

Noël, L. 1993. *Intolerance: A general survey.* Kingston and Montreal: McGill-Queen's University Press.

Royal Commission on Aboriginal Peoples (RCAP). 1996. *Report of the Royal Commission on Aboriginal Peoples.* 5 vols. Ottawa: Canada Communication Group.

Smith, G. 2000. Protecting and respecting Indigenous knowledge. In *Reclaiming Indigenous voice and vision*, ed. M. Battiste, 209–24. Vancouver: UBC Press.

Smith, L. T. 1999. *Decolonizing methodologies: Indigenous peoples and research*. London: Zed Books.

– 2000. Kaupapa Maori research. In *Reclaiming Indigenous voice and vision*, ed. M. Battiste, 225–47. Vancouver: UBC Press.

United Nations. 1989. *Report of the United Nations seminar on the effects of racisim and racial discriminaton on the social and economic relations between Indigenous peoples and states.* UN Commission on Human Rights, 45th Sess., UN Doc. E/CN.4/1989/22.

16

The Anishinaabe Teaching Wand and Holistic Education

ROBIN CAVANAGH

INTRODUCTION

Twelve years ago, I met an Ojibway Elder whose influence would forever change my life. At that time, I began to walk the path my mother had left over 50 years before when she was forced to enter Residential school – a path full of people, like myself, struggling to regain a culture and a tradition half lost within a contemporary world.

I have spent many years with Elders, attending ceremonies, fasting, and sitting with the Ojibway drum, in order that I might build a good Creational-view as I journey through life. It is through this Creational-view imparted to me by the Elders, along with the knowledge and experience I have gained in the last eight years as a university student, that I have pulled together what I consider to be an Aboriginal Educational Framework.

The purpose of the framework is to highlight aspects of an Aboriginal Creational-view of education that expands the often-limited interpretation of the term "holistic." As a framework, it demonstrates the complexity of traditional Aboriginal knowledge systems; at the same time, it reveals a process that may be used today to assist Aboriginal peoples in developing programs, curricula, and human resources that are culturally specific and complementary to [and exceeding] Western standards of education. It also provides an excellent model for others exploring alternative methods in education.

At a time when Aboriginal peoples are gaining more and more control over their educational institutions, programs, and curricula, and have begun addressing their unique cultural needs within other institutions, discussions relating to processes of Aboriginal education become key.[1] I offer this work, not as a blueprint, but merely as a point of reference, to contribute to a discourse and process that begins from within Aboriginal communities and includes its members, educators, Elders and others who have been invited to participate. Participatory involvement is critical to developing programs and curricula that are relevant to the unique experience, cultural ways of knowing, and environment of a given community.

Aboriginal community environments have changed dramatically since our original teachings were received. Despite this, traditional methodologies can be as relevant today as they ever were (Boldt 1993). Personal experience with Elders' teachings has revealed that, as lessons, they hold no definitive answers. Instead each embodies a process or way of "being and becoming."[2] Today, these traditional methods offer great potential for the development of educational programs and curricula for Aboriginal Peoples.

For traditional teachings and methodologies to assume any relevance, however, we must relearn, understand, and apply them to re-evaluate the world around us. They can not only provide a foundation for educational programs and curricula, but also guide us through the many choices that, as Aboriginal peoples, we must make. Found within traditional stories, legends, and teachings, these processes are also fundamental to the development of individual identity and a way of being and becoming within our current educational environment.

The traditional Aboriginal approach to education was premised on the value of developing creative, responsible, and balanced individuals equipped to handle the challenges they would face within Creation. It required the commitment of teachers, family, community and Nation, and was responsive to the student and her educational desires and needs (Johnston 1976). The student's gifts, once realized through participation in ceremonies, educational activities, and community life, provided the foundation for further training. Traditionally, new teachings were imparted to the student when she was ready, that is, when the student had achieved certain levels of understanding. Instruction continued at a pace that was respectful of the student's ability and needs, while also challenging the student's creative capacity. The educational process was student-centred and balanced by the recognition and participation of all of Creation.

The pages that follow provide an example of this educational process by considering the traditional life journey, from birth, of an individual within her community. The Teaching Wand provides the framework within which some of the essential values, beliefs, and processes are highlighted to describe the individual's education and progression through the various stages of life.

The elements presented in the Wand are drawn from various Aboriginal sources, and the many teachings I have received from Elders. I came across the structure of the Ojibway Teaching Wand itself five years ago in an article by J.W.E. Newbery in which he describes its use by an Elder in ceremony (Newbery 1977).[3] Its application as a framework from which to examine the complexity of traditional methodologies and values as twelve figures laid one over the other emerges from my attempts to characterize the Aboriginal meaning of the term *holistic* (Cavanagh 2000).

Many of the values presented within the figures of the Teaching Wand already had specific directions associated with them, according to oral teachings. Others did not. I have positioned the latter with respect to their meaning, the flow of the Teaching Wand, and the teachings I have received. The *placement* of these values and processes – with certain directions – is not absolute. The importance lies in their *meaning*. Individuals and/or communities may place them differently depending on their own cultural teachings and understanding.

It should be noted that many Aboriginal peoples are currently struggling to relearn what remains of traditional culture from the few Elders that still have the knowledge. Most Aboriginal peoples today share a common vision with their ancestors – to experience a *good life in a good way*. For many, the vision begins with renaming and reclaiming the world we live in today, based on traditional ways of knowing and explaining the world around us. Elders are doing their part by sharing the teachings they have. As an individual, it is my responsibility to apply these teachings to the environment I now find myself in as I begin to re-envision the world around me.[4]

The process of renaming the world requires that Aboriginal peoples take a close look at themselves as well as the world around them. Waynaboozhoo, in his task of naming the world, recognized the good and the bad in all things. His honesty allowed him to name the world truthfully (Benton-Banai 1988). The process of being and becoming, surviving and prospering, as Aboriginal peoples requires that we accomplish the same in our renaming of the contemporary world.

A critical and honest analysis is never easy. A great amount of documentation already exists regarding the unhealthy conditions that

Aboriginal peoples find themselves in today (RCAP 1997). Many of the causes of these conditions have also been identified, and some solutions have been put forward. The question remains whether the proposed solutions are based on *our own* renaming of the world, or on the assessment processes of others. With efforts underway within Aboriginal communities to restore the health of the people (RCAP 1997), it becomes critical that any solution/methodology put forward should reflect our own understanding of the world. Without this, we are forced to acquiesce to another's definition of what it means to be a healthy Aboriginal. Given that our Nations are able to create the environment necessary to become healthy, it follows that Aboriginal communities and institutions will require a healthy Aboriginal education system from which to draw support.

Creating relevant educational programs and curricula requires that, at the outset, Aboriginal peoples take a step away from the current system to provide the space to rename and reclaim their own unique approach. We must recreate an Aboriginal education system founded on traditional values ensuring cultural continuity throughout. The option then exists to expand on and/or include other existing methodologies.

An Aboriginal education system would strengthen our children's sense of self and place, and would provide them with the opportunity to apply their knowledge to create choices for their future that reflect their rich cultural heritage. It would contribute to a positive Aboriginal identity within the students and Nations in general. As a wise Elder once said, "The people will always know what to do provided they have the context in which to do it" (Solomon 1990, 171). The traditional education system proposed herein is spirit-centred, Creationally viewed, and process-oriented, within a holistic system operating on a reciprocal relationship with Creation (Cavanagh 2000).

SEVEN STAGES OF LIFE

Traditional Aboriginal peoples believed that they entered Creation as mind, body, and spirit (Thrasher 1987). The individual upon entering her community was considered sacred. The community understood that its responsibility was to provide nourishment and education for mind, body, and spirit in a spiritual, emotional, physical, and mental way (Bopp et al. 1984). The individual reciprocated by adding to the community's potential to survive, prosper, and create. By understanding, actualizing, developing, and sharing the gifts provided to the individual by the Creator, she was fulfilling her responsibilities within Creation.

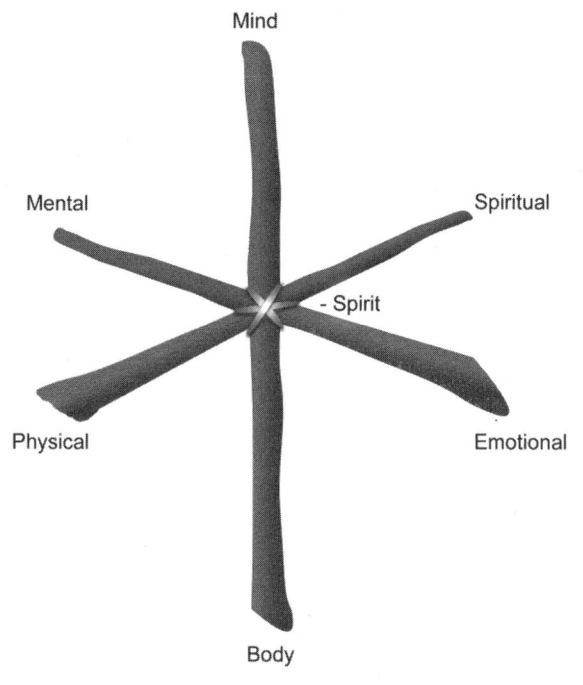

```
            Mind
             |
Mental       |       Spiritual
     \       |       /
      \      |      /
       \     |     /
        \    |    /
         \   |   /
          \  |  /
           \ | /
            \|/— Spirit
            /|\
           / | \
          /  |  \
         /   |   \
        /    |    \
       /     |     \
      /      |      \
     /       |       \
Physical     |      Emotional
             |
             |
            Body
```

Figure 1   Three Levels of Learning

Figure 1 illustrates the human make-up and potential of a student, as well as the three levels of learning – mind, body, and spirit. For the learning process to occur in a healthy way, each of the three aspects must be nourished and guided by the educator working with the student. The educational relationship between the new student and her community traditionally followed a long and very thorough process guided by the concept of the Seven Stages of life (Manitowawbi as quoted by Rheault 1999). In the first stage (Figure 2), the community assumed the responsibility of providing for and educating the individual. Stories, legends, and song were used to convey the Creation Story, history, and original instructions of Creation. The new student of Creation was introduced to the educational process by some of the wisest and most respected of the community's teachers, the Elders.

During the period of *fast life,* the individual's knowledge expanded to include possible roles within the community. The student also developed her sense of choice and responsibility toward her growing relationship with Creation. It was during the latter part of this stage, or early within the next stage – *early adulthood* – that a vision was sought during the first

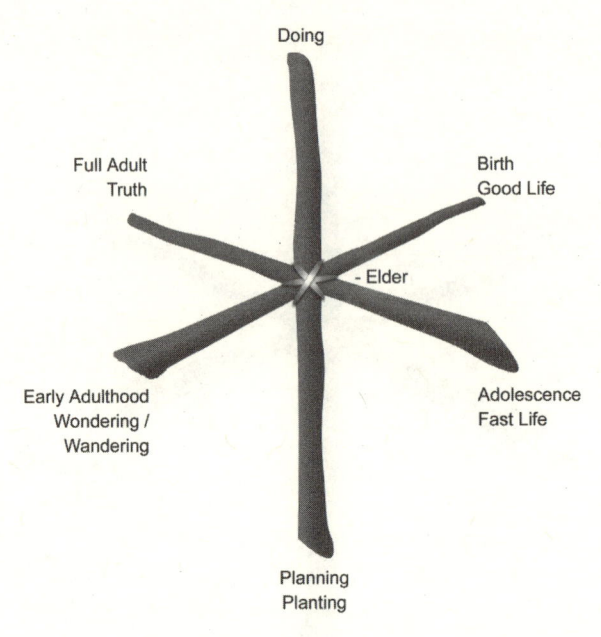

```
                    Doing

    Full Adult              Birth
    Truth                   Good Life

                    — Elder

    Early Adulthood         Adolescence
    Wondering /             Fast Life
    Wandering

                    Planning
                    Planting
```

Figure 2   Seven Stages of Life

Fast. The sacred process of fasting aided the student in understanding her unique gifts and gave her a greater understanding of Creation itself. Through the spiritual guidance of Elders, Traditional Teachers, and with the support of the rest of Creation, the individual sought to understand and actualize her vision. This was one of the most empowering periods for the Aboriginal student, seeking to understand her role within Creation. Throughout this period, the possibilities that existed within the community were shared with, and explored by, the young adult. She experienced many challenges as part of her education to find her place and purpose. The community supported the individual, but ultimately it was the individual who was responsible for understanding and actualizing her gift and potential.

During the stage of *truth*, the student began to accept full responsibilities as an adult. Only through the actualization of her vision and the utilization of her gift did the individual fulfill her potential as a "good person." The community benefited as the individual participated in the process of creation by creating and learning while she walked through life.

The *planning and planting* stage involved raising families, becoming teachers, and beginning the process of re-creation within the commu-

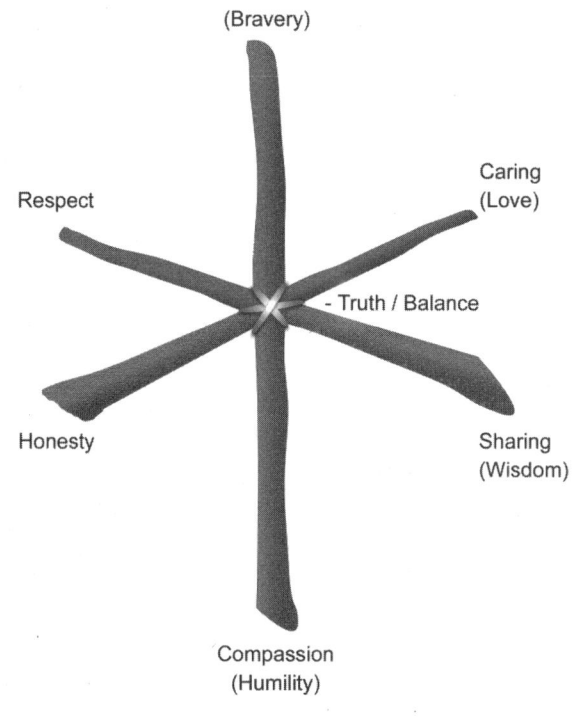

Figure 3   Seven Laws/Gifts

- Bravery (top)
- Caring (Love)
- Respect
- Truth / Balance (center)
- Honesty
- Sharing (Wisdom)
- Compassion (Humility)

nity. As she received, from her earliest time within the community, the gift of life through her birth, the individual gave back to Creation one of its most sacred assets – new life. In so doing the individual fulfilled her original instructions, and continued to renew the relationship with Creation.

The cycle of Creation and education continued. During the stage of *doing and rediscovery*, as her children grew, the individual began to reflect upon her experiences and knowledge gained within the community. It was a time of rebalancing herself within Creation, and gaining a deeper understanding of Creation and her own responsibility within it.

In the final stage of life, *elder/traditional teacher*, the individual accepted the responsibility of sharing her experience and understanding of Creation with the community. Through teaching, guiding, and storytelling, the process of Creation was passed on; each stage of life unfolding into the next and all occurring simultaneously, without beginning or end.

SEVEN LAWS/GIFTS

Figures 1 and 2 demonstrate how one might trace an individual's development as she progresses through the seven stages of life. Moving on to Figure 3, which depicts the seven laws/gifts, we begin to see the development of a framework of processes guiding Aboriginal education.

The traditional Anishinaabe were given seven laws or gifts to guide them as they journeyed through the seven stages of life (Johnston 1976). They needed to come to understand the values and theories of these laws, and to apply them in a balanced way. These values, beliefs, and theories were an integral part of the educational relationship between the individual and the rest of Creation. The student journeying through the seven stages of life was considered a sacred part of Creation, and in return, she treated all of Creation in a sacred way.

Each of the seven laws contained its opposite or dark side within it (Nabigon 1999). If not shared and understood in a balanced way, the possibility of harm existed for both the student and the community. This educational theory promoted the responsibility of utilizing knowledge in a truthful way, which meant in balanced consideration of all laws/gifts. The theory was premised on the idea that one law/gift in itself could be harmful if not used in balanced consideration of the others – as an indication of choice and decision-making within the community.

The concept of balance is key. As noted earlier, an individual's human make-up comprises mind, body, and spirit. The traditional student received teachings and experienced Creation in a spiritual, emotional, physical, and mental way. These aspects were developed as the individual followed the seven stages of life. The seven laws/gifts were imparted to Aboriginal people in varying degrees of complexity as they progressed through various stages of life; continually nourishing mind, body, and spirit. In return, traditional Aboriginal students applied their knowledge to participate within the community at all levels of responsibility.

Areas of Responsibility

The responsibilities of the individual (Figure 4) affirmed an in-depth understanding of Creation that she developed as she progressed through the seven stages of life. Many Aboriginal peoples were taught that, as a sacred part of Creation, they were individuals who were part of a family that had two Clans within it which, together with other families, belonged to a community. The community was also part of the Nation,

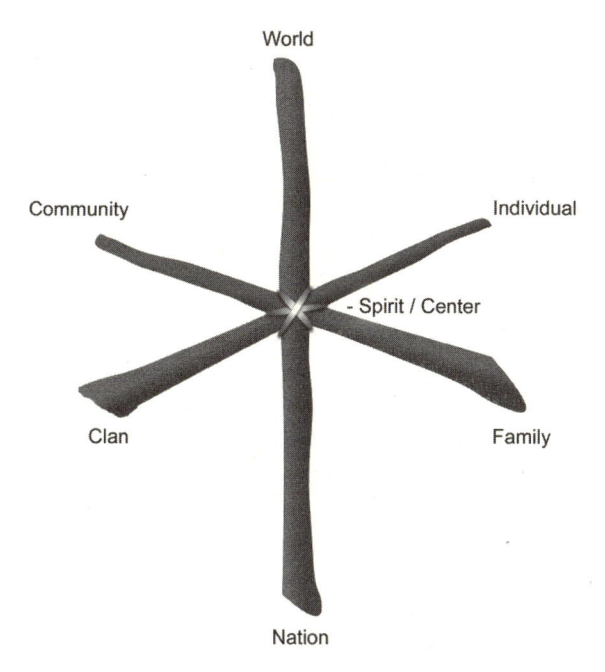

Figure 4   Responsibilities

which was a sacred member of the world that was made up of many different Peoples. At each stage the individual's knowledge of, and experience with, the larger human family expanded.

The individual's responsibilities and choices within this human family were assumed in a balanced way, as all aspects were considered sacred. Although each individual ultimately had freedom of choice in life, having received and understood the appropriate teachings, she recognized that her choice had an effect on the rest of Creation. The student developed an understanding of these areas of responsibility as her education progressed through the seven stages of life. At each stage, the depth and understanding of the responsibilities and choices one had within this human family became clearer.

The Seven Directions

The Aboriginal student's education was also guided by the teachings of the *Seven Directions* (Figure 5). Each of the seven directions contains teachings about both the individual and the community. The process implied by the teachings of the seven directions is reflected in the sacred-

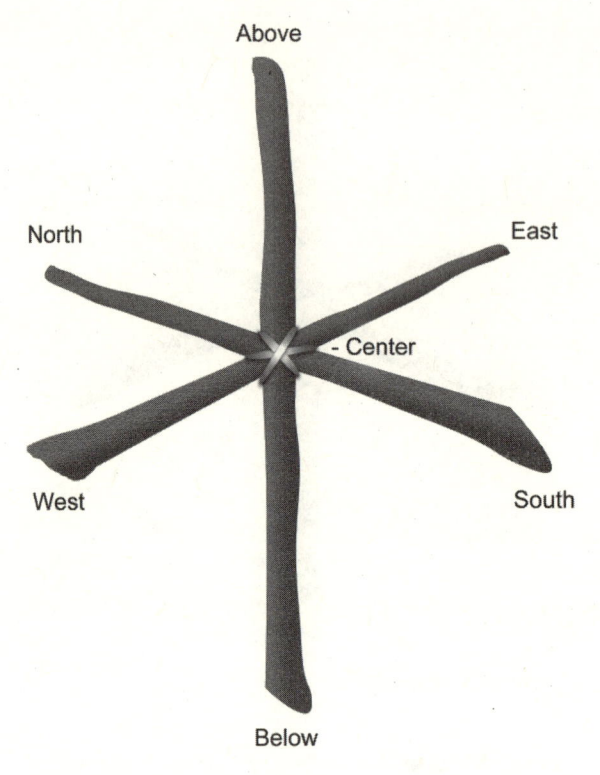

Figure 5 Seven Directions

ness and interconnectedness of the individual with the rest of Creation. As her education proceeded, and the further the student explored a given direction, the deeper her understanding became of the lessons found in that direction, and the closer the connection became between that individual and Creation. It was inherent to the educational process that an individual journey into each of the directions, returning eventually to the centre to reflect upon the knowledge she received. In this way, the community benefited from the sharing of knowledge by the individual who was following her ideal path through the seven directions.

The depth of knowledge gained within each direction was correlated to the stage of life of the individual. The traditional Aboriginal student went through this educational cycle many times in a lifetime. Each time her understanding deepened. The depths of understanding found in the gifts/teachings of the seven directions were recognizable to those individuals ready to receive them.

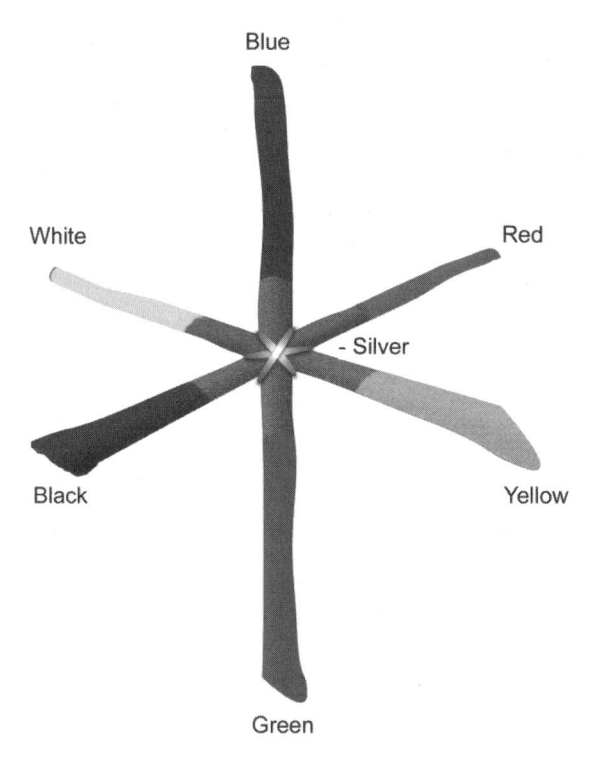

Figure 6   Seven Colours/Four Races

Seven Colours/Four Races

The focus thus far has been on the individual and/or some of the processes used in the education of the individual within the community. The teaching of the *Seven Colours/Four Races* (Figure 6) also played a significant role in education, and reiterated many of the teachings of the seven directions. One of these is the teaching of *the four races* of humans that appears within the Anishinaabe Creation story. The four colours represent the four distinct Peoples that are recognized as having received their original instructions from the Creator.[5]

As sub-parts of Creation, each of the four races had responsibilities to fulfill according to their own teachings. It was understood that all of the races entered Creation with mind, body, and spirit and each was to experience their part of Creation in a spiritual, emotional, physical, and mental way.

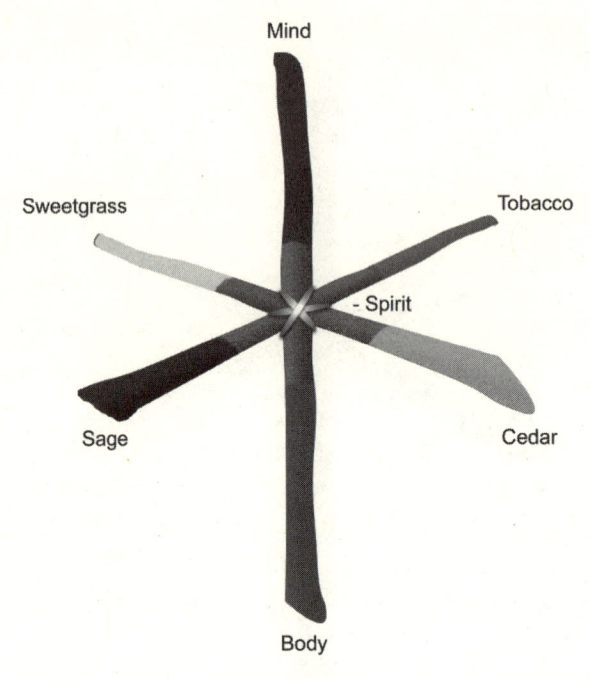

Figure 7 Plants and Medicines

The significance of the above understanding to this theoretical framework is that the Anishinaabe recognized the existence of their human brothers and sisters within Creation as sacred beings, and respected their brothers' and sisters' choices and interpretation of their original instructions, and expected the same respect in return.

Plants/Medicines

The plant world, which existed before animals, was a sacred part of Creation vital to the educational processes of Anishinaabe peoples and provided many gifts and teachings (Figure 7). The plant world sustained the lives of the animals and people and was honoured and respected.

The concept of reciprocity was imparted to Aboriginal students at a very early age, first through stories, songs, and ceremony, and then through observation and in-depth teachings as they grew within the community. Every living element (human or otherwise) within this system had its place – original instructions – and all parts were essential to the continual process of Creation.

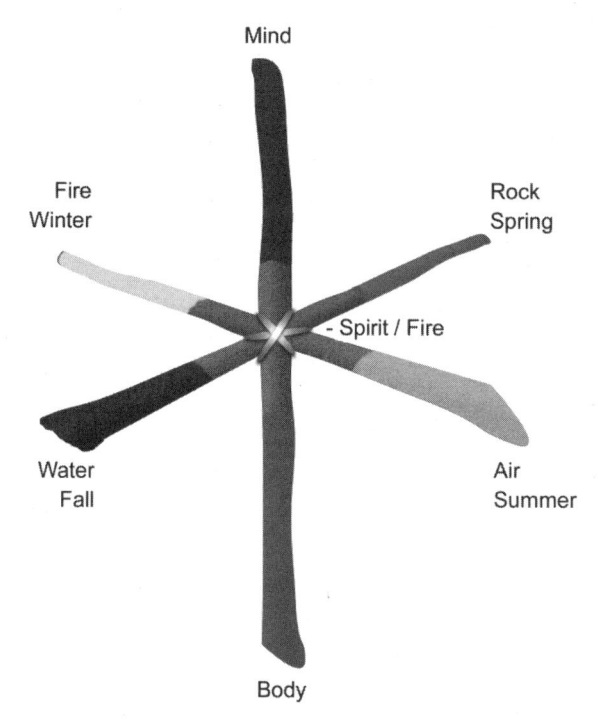

```
              Mind

  Fire                    Rock
  Winter                  Spring

              — Spirit / Fire

  Water                   Air
  Fall                    Summer

              Body
```

Figure 8   The Elements

Students learned of the sacrifices made by the animals and plants and honoured them within ceremony, stories, and songs. In fact, many Aboriginal peoples communicated and sent their prayers, and thanks, to the Creator via these "older brothers and sisters." In many cases, the animals were considered facilitators of change, carrying messages from the Creator through dreams and visions.

The recognition and reverence accorded to the plant and animal world, presented as an educational process within the teaching wand, demonstrates the interrelatedness of traditional methodologies and how all of Creation was recognized and respected by Aboriginal peoples.[6] Traditional Aboriginal educational processes represented a balanced system premised on core values and a belief in reciprocal recognition that all Creation was sacred and interdependent. Through the Creator, the breath of life runs through the entire system. And it is this breath of life that is at the core of everything within Creation (Johnston 1976). Aboriginal students learned and understood that the spirit instilled in all of the interrelated parts of Creation was Creator manifest.

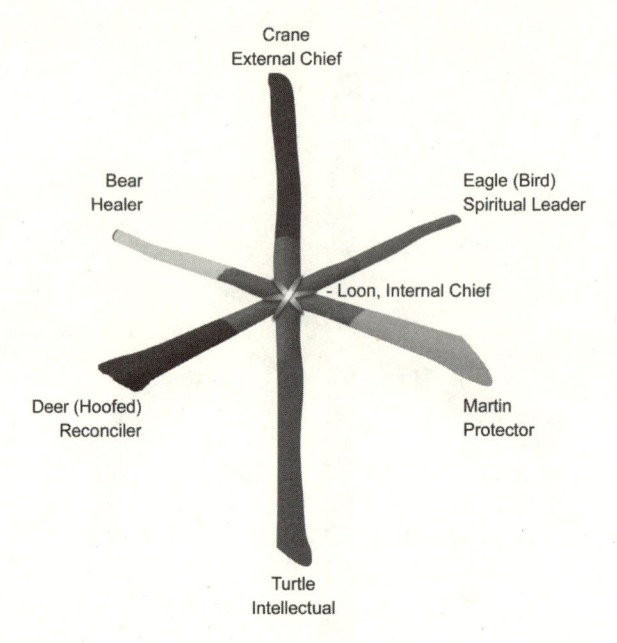

```
            Crane
         External Chief

  Bear                    Eagle (Bird)
  Healer                  Spiritual Leader

              — Loon, Internal Chief

  Deer (Hoofed)            Martin
  Reconciler               Protector

            Turtle
         Intellectual
```

Figure 9   The Clan System

The Elements

Aboriginal peoples hold a great reverence for Mother Earth. As the physical space that life was actualized upon, she has provided Aboriginal peoples a place within Creation to carry out their original instructions and continue their relationship with Creation. Mother Earth is composed of four primary elements – earth, air, water, and fire. The Creator instilled in each the breath of life (Figure 8). These four sacred and essential elements provided the foundation from which all of Creation was born, including Aboriginal peoples, and without which Creation would not be (Johnston 1976).

The Clan System

During a time of great difficulty, the traditional Anishinaabe received the gift of the Clan system (Figure 9). Many Creation stories state that spirit beings, sent by the Creator, facilitated the introduction of the Clan system. More Clans emerged as the Anishinaabe grew in number and spread across Turtle Island. Guided by their original instructions and understanding of the Clan system, new Anishinaabe Clans were required to represent the values deemed necessary for the education and continuation of this governing system (Benton-Banai 1988).

Figure 10   Grandmother Moon

The Clan system did not represent a separate governing body making decisions for the people. Rather, it represented a methodology by which the community could ensure that it utilized to the best of its ability all its resources to inform, to evaluate, to debate, and to create respectful decisions on any issue brought before the Council. The individuals chosen to represent their families, Clans, communities, Nations, and Creation were those who demonstrated the best abilities and gifts to do so. From birth, the Anishinaabe were prepared and educated for their roles and responsibilities within the Clan system. The Clan system was a sub-part of Creation through which the Anishinaabe resolved issues and made choices pertaining to Anishinaabe internal and external responsibilities within Creation.

Grandmother Moon

Female teachings, changing seasons, and cycles within Creation are represented by Grandmother Moon (Benton-Banai 1988). The teachings of Grandmother Moon are represented with a circle in the Teaching Wand to honour all that Nokomis gave and continues to give to Creation (Figure 10). Change within communities was (and is) continuous. However, within change there was also constancy. Aboriginal educational practices

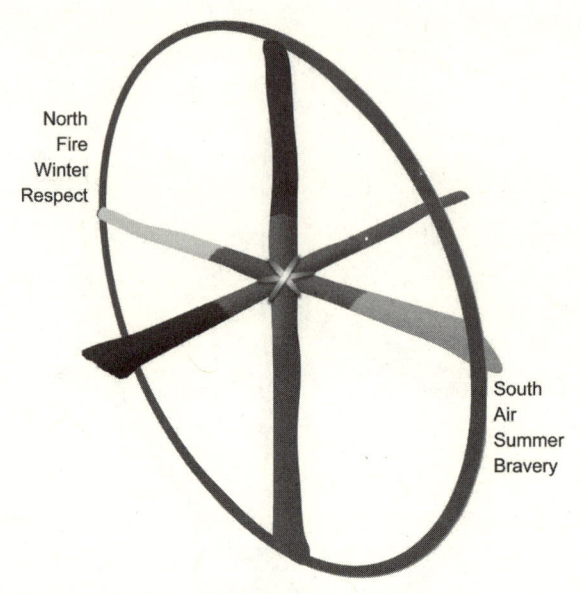

```
       North
       Fire
       Winter
       Respect




                                South
                                Air
                                Summer
                                Bravery
```

Figure 11   Grandfather Sun

ensured, during times of change and new experience, that the methodology used to guide decisions – despite changes in the environment – remained constant (Benton-Banai 1988). Thus, as Aboriginal peoples experienced changes in their environment, their specific roles and responsibilities were adapted to meet the needs and choices of all community members.

As Grandmother Moon remains constant in her position within the night sky and sheds light on many aspects of Creation, so too does Grandfather Sun.

Grandfather Sun

New ideas entering the community are given illumination and energy to grow from Grandfather Sun (Benton-Banai 1988). Grandfather Sun (also depicted as a circle around the Teaching Wand) represents male teachings and the spirit fire within all living things (Figure 11). His significance to Aboriginal educational methodologies is that all life within the community needs energy to sustain itself. Energy is essential to the creative process within the community. The spirit fire gives life and purpose to the community. The need to nourish the spirit fire drives the creative search for answers within the Great Mystery.

The Anishinaabe Teaching Wand

Blue, Sky, Honour, Bravery, Crane, Above, Nation, Mind, External Chief

White, North, Mental, Fire, Winter, Sweetgrass, Respect, Community, Healer, Bear

Red, East, Earth, Eagle, Spring, Caring, Love, Individual, Spiritual Leader, Tobacco

Balance – Silver
Spirit | Truth

Black, West, Physical, Water, Honesty, Fall, Reconciler, Deer, Honesty, Clan, Sage

Yellow, South, Air, Emotional, Summer, Cedar, Sharing, Family, Martin, Wisdom, Protector

Green, Body, Earth, Below, Compassion, Humility, World, Turtle, Intellectual

Figure 12   Creator/Creation

The creative search to understand life's mysteries is tempered by the constancy of Grandmother Moon, and by the understanding that all aspects of life must be balanced according to our original instructions and relationship with Creation.

The concepts of Grandmother Moon and Grandfather Sun permeated educational methodologies, teaching traditional Aboriginal peoples a theory of relationships within the community. Two circles, signifying moon and sun, are placed within the Teaching Wand, visually demonstrating the importance of both male and female teachings, knowledge, understanding, and choice. Just as essential is the recognition that the two circles intercept, representing the understanding that these teachings should be considered within a relationship of respect for each other in their given Creational contexts. As environments changed, the day-to-day roles and responsibilities of men and women also changed, yet each remained guided by the values inherent to the Aboriginal Creational View.

Creation/Creator

A ring connecting the four sacred directions is placed in the last figure of the Teaching Wand to represent the Creator/Creation (Figure 12). With the final circle in place, a sphere is created, representing a teaching methodology premised on the core values and beliefs of the traditional Aboriginal Creational view. Thus, we are reminded that each educational process returns to the beginning of Creation, or spirit. Spirit is instilled in all of Creation; Spirit is Creation's facilitator. And the process completes itself in spirit, and in the creation of humans. Again, we remind ourselves that a traditional Aboriginal educational framework is spirit-centred, Creationally viewed, and process-oriented within a holistic system operating on a reciprocal relationship within Creation.

The framework is completed with no beginning and no end, presenting a spherical Creational view of education. The cycle never ends. Aboriginal children entered Creation with mind, body, and spirit; Nokomis nourished this new life in a spiritual, mental, physical, and emotional way. Through story and song, Nokomis instilled within the child all that she was. In return, Nokomis received all of Creation in her arms; and the great mystery continued.

CONCLUSION

The framework/educational theory discussed in this paper outlines the creational processes of education found within the many teachings of Aboriginal peoples. The methodology is as relevant today as it was in the past. It is a given, of course, that the goal of education is to foster in Aboriginal students a sense of identity and pride based on their complex and unique cultural foundation. The methodologies, or ways of being and becoming, vary according to the many teachings found within Aboriginal communities and Nations and to the particular environmental contexts in which Aboriginal peoples find themselves.

"Holistic" processes, from the perspective of an Aboriginal Creational view, are complex and thorough, and can be used to develop, evaluate, and sustain Aboriginal programs and curricula in any environment. A traditional foundation instills students with pride. Equipped with the knowledge of their ancestors, they are capable of meeting any challenge life has to offer them in their journey to re-name and reclaim their place within Creation. Aboriginal students can only benefit from an educational methodology that provides them with the theories and processes necessary to understand, evaluate, and take action based on an Aboriginal way of knowing and making decisions.

The Elders continue to understand their role and responsibilities within our communities and continue to teach those who are ready and willing to learn. "How can we expect to answer 'What is the meaning of life?' when the other half of the question is not asked – 'What is the purpose of life?' There is a purpose and it is sacred and precious. We have to understand that we have the opportunity, responsibility and the power to change the world. The whole system told us we are powerless. It is not true" (Solomon 1990, 102).

It is the responsibility of students, teachers, communities, and nations, in the appropriate stages of development, to listen, learn, understand, and actualize what has been shared by the Elders, and to use these processes and teachings in the re-creating and re-claiming of our educational system. The framework and educational methodology described herein have provided only an outline. The precise teachings, stories, and songs are held in the hearts and minds of the Elders and traditional teachers who have dedicated their lives to the continuance of this knowledge.

NOTES

1 "The chiefs of the Nishnawbe-Aski Nation recognize education as fundamental to the maintenance of [their] Nishnawbe identity and well-being, as well as to the development of [their] communities (NAN Website)."

2 "Being and Becoming is primarily concerned with the process of the individual's being and becoming a unique person, responsible for his or her own life and actions in the context of significant group situations. The traditional Native being-becoming posture requires trust of self and others, a non-manipulative relatedness and a sense of oneness with all dimensions of the environment, components which, without exception, are experienced and perceived as possessing a life energy of their own. Native philosophy of life manifests characteristic person centeredness, a holistic personalism that regards the human person as a subject in relationships: both the subjects and the relationships exist in a dynamic process of being-becoming" (Couture 1985).

3 It is apparent in Newbery's article that the only limitations to what *could* be taught in respect to a traditional Creational-View were the limitations of the

teacher using the Teaching Wand. Thus I took the liberty of exploring its use in my Major Paper at York University.

4 See Couture's article "Traditional Native Thinking, Feeling, and Learning" (Couture 1985, 6) for a discussion of values and beliefs found within Aboriginal oral teachings. Couture contends that Aboriginal people survive by upholding these values and beliefs, and that although these teachings are passed on through the Elders, the interpretation and application of these teachings is the responsibility of the younger generation (Couture 1985, 6).

5 In response to a reviewer's concern that this may represent essentialism, the author explained, "The four colours are symbolically representative of the Nations and peoples of the world. They are used to refer to the differences that exist throughout our Earth, with an emphasis on honoring difference while respecting the uniqueness of individual/community/national experience and ways of being" (Editors' note, P.T. and L.M.).

6 There are four orders in creation. First is the physical world; second, the plant world; third, the animal; last, the human world. All four parts are so intertwined they make up one life and one whole existence. With fewer than the four orders, life and being are incomplete and unintelligible. No one portion is self-sufficient or complete; rather, each derives its meaning from and fulfills its function and purpose within the whole of creation (Johnston, 1976: 21).

REFERENCES

Benton-Banai, E. 1988. *The Mishomos book*. Wisconsin: Indian Country Communications, Inc.

Boldt, M. 1993. *Surviving as Indians*. Toronto: University of Toronto Press.

Bopp et al. 1994. *The sacred tree*. Alberta: Four Worlds Press.

Cavanagh, R. 2000. *Towards an Anishinaabe organizational theory*. Unpublished paper, York University.

Couture, J. *The role of Native Elders: Emergent issues*. unknown source, unknown publication.

– 1985. *Traditional Native thinking, feeling, and learning*. Unknown source.

Johnston, B. 1976. *Ojibway heritage*. Toronto: McClelland and Stewart.

Nabigon, H. 1999. *The hollow tree*. Manuscript: Unpublished.

Newbery, J.W. 1977. *The universe at prayer*. In *Native religious traditions*. Waterloo: Wilfrid Laurier University Press.

RCAP (Royal Commission on Aboriginal Peoples). 1996. *Report of the Royal*

Commission on Aboriginal peoples. 5 vols. Ottawa: Canada Communication Group.

Rheault, D'A.I. 1999. *Anishinaabe Mino-Bimaadiziwin (The way of a good life)."* MA thesis, Trent University.

Solomon, A, 1990. *People songs for the people: Teachings on the natural way.* Toronto: NC Press Limited.

Thrasher, M. 1987. *Life decision.* Michael Thrasher.

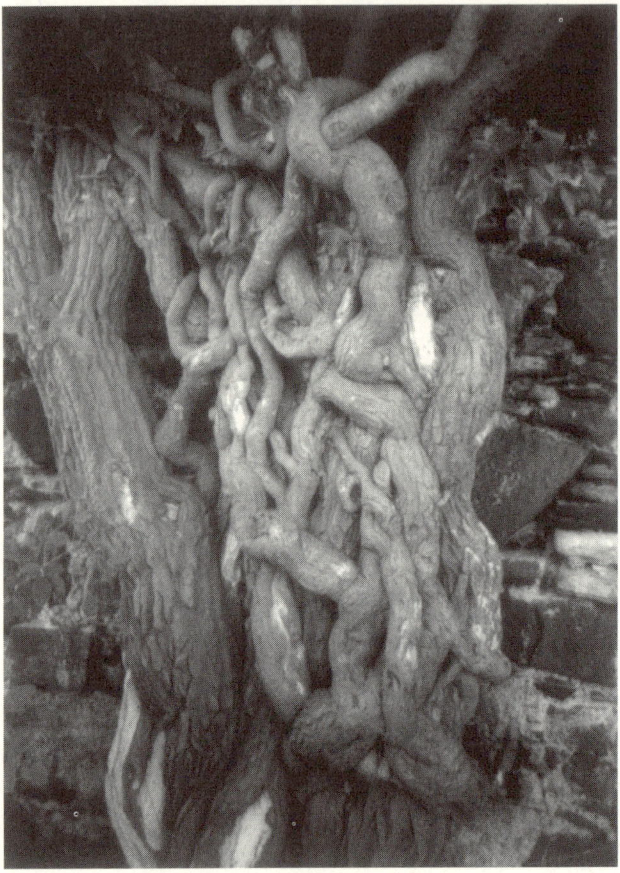

17

Visions for Embodiment in Technoscience

NATASHA MYERS

MAKING VISIBLE THE INVISIBLE

Biology is an intensely visual practice: vision is the primary sense of the biologist whose occupation is observation. In the Western culture of science, vision is considered more objective than the other senses: "seeing is believing," eyes are supposed to be "windows to the world." In this positivist idealization of objectivity, to render visible is to provide proof, and so biologists use an extensive array of visualization technologies and modes of representation, including microscopes, cameras, chemical reagents, genetic markers, computational models, and multi-dimensional imaging tools, to translate the things we can't see into visible evidence (Latour 1990, Lynch 1990, Amann and Knorr-Cetina 1988, Keller 1996 and 2002). These technologies make visible otherwise invisible life processes, and rapid advances in biological imaging strategies have made possible otherwise impossible views of the biological world. Yet do these tools generate lucid, objective translations of a clearly delineated biological reality? If not, what are the limits and contingencies of vision in biological technoscience, and how do these limits frame biological knowledge?

Environmental philosopher Neil Evernden suggests that a widespread faith in Cartesian vision gives biologists the "luxurious delusion" that they are "neutral observers" with an objective perspective on the world (1985). Embedded within a moral economy of "mechanical objectivity" that defers to the presumed dispassionate authority of "machine vision," biological observation is made to seem like passive and detached voyeurism (see Daston and Galison 1992, Galison 1998). The elaborate technologies biologists use to render visible the intricate mechanisms of life are commonly perceived as means for distancing the scientist's body and all its complicating agency, subjectivity and values; an attempt to erase the observer from the scene. Feminist science scholar Donna Haraway identifies this false promise of objectivity as a "myth of body-lessness" (2001). We wield our eyes like "body-less" instruments assuming that they grant impartial access to knowledge of the world.

Idealizing a disembodied rationality, scientific reports tend to obscure the agencies and interests of the biologists who produce them. Writing themselves out of the picture, biologists disappear the multiple subjectivities and agencies (including culture and politics) embedded in the materials and methods of their practices. Obscured by a complex set of ideologies, the subjectivities of the organisms that biologists study are also made to disappear in the process. Drawing attention to the ways that power, knowledge, and vision are intertwined in Western narratives of human 'dominion' over nature, this paper explores what is at stake in the ways that biologists lay claim to truths.

In its manipulation of living organisms, biological visualization often becomes a violent and invasive practice. "Vision," says Haraway, "is always a question of the power to see – and perhaps of the *violence* implicit in our visualizing practices." She asks: "With whose blood were my eyes crafted?" (1991, 192) The biologists' eyes, as their "windows" to knowledge, are stitched together from the fragmented flesh of once-living organisms. Blood is almost always spilled in attempts to make visible the inner workings of otherwise invisible lives. Is biological vision necessarily a violent practice? If so, how might issues of ethics be addressed within biological imaging practices?

In her desire to transform scientific objectivity by holding scientists accountable for the ways that they render the world visible, Haraway holds out for the promise of perspectives that can offer "knowledge potent for constructing worlds less organized by axes of domination" (1991, 192). The illusion of disembodiment tends to forestall scientists' accountability for their participation in practices of visual production. Haraway suggests that recognizing the "body-fullness" (2001) of vision in science might be a means to implicate biologists as co-constructing agents in knowledge production. Holding biologists accountable for the intent and consequences of their actions is an attempt to integrate "situated knowledges" as an embodied ethic into biological practices (Haraway 1991).

Maurice Merleau-Ponty (1962, 1968) also offers insight into the ways that bodies are implicated in seeing. Providing a phenomenological practice for improving our perceptual skills, his embodied approach to vision meets Haraway's call for "better accounts of the world" (1991) and more "modest witnessing" (1997). Both Haraway and Merleau-Ponty challenge the Cartesian notion of disembodied vision by demonstrating the participation of both observer and observed in processes of visual production. Both scholars attest that vision might be reclaimed as an embodied and situated practice, offering indispensable insight into the possibilities for embodied vision and embodied ethics in biology. I investigate a potential cross-hybridization between their theories of embodiment for a practice of biological imaging that does not strip human or nonhuman lives of their agencies, subjectivities, or animate modes of "being-in-the-world."

How might biological imaging technologies and practices engage our bodies, rather than alienate them? How might "embodied" vision generate more complete (i.e., less reductionist, less mechanistic), more "living" and "livable" knowledges of the world? Here I explore how "body-full" vision might contribute to an ethic for visualization practices in biology.

THE FLESH AS AN ETHIC FOR BIOLOGICAL VISION

In *The Visible and the Invisible* (1968), Merleau-Ponty develops a theory of "the flesh" to describe the phenomenon of vision. The flesh is a useful metaphor for understanding visualization practices in biology because it accounts for the dual embodiment and intimate relationality that inheres between all kinds of bodies caught up in acts of seeing. Describing a "reversibility" between seer and seen, Merleau-Ponty suggests that things and beings slip and slide in and out of their commonly designated orders of "subject" or "object" (see Merleau-Ponty 1968). As such, the flesh is evocative of the inter-implication of self and world, suggesting ways of re-conceiving and re-figuring the relationships between subject and object, mind and body, and seer and seen. As a "body-full" metaphor for ways of seeing, the "flesh" can be extended to convey the full agency and participation of all bodies (human, nonhuman, and machine) engaged in practices of biological technoscience.

More than this "reversibility" between subjects and objects, Merleau-Ponty suggests that vision crosses over into touch, and touch into vision. Drawing on Judith Butler's interpretation of flesh, Gail Weiss describes this as a kind of inter-sensory "'transubstantiation' of vision into touch, movement into expression, whereby I see by 'touching' and move by 'speaking' with my body" (Weiss 1999, 119). Rather than conceiving of eyes as disembodied instruments of vision, sight and seeing become the capacities of whole bodies, fully taken up in acts of perception. In this "crossing over" between the "visible" and the "tangible," our visual exchanges with others engage us viscerally in our encounters with the world (Merleau-Ponty 1968). In this way, our hands can be recognized as fleshy extensions of our eyes, so that looking becomes a way of feeling out the world with our eyes, and seeing, a way of being touched by others. Such a transubstantiation of vision into touch, of one sense into the other (a sort of synaesthesia), is integral to Merleau-Ponty's understanding of the embodiment of all perception. Vision is thus transformed from an objectifying into an "embodied" practice (1968). Significantly, by drawing vision into the body, the flesh displaces assumptions of the primacy of vision over other bodily senses, and opens up a space for re-imagining the perceptive capacities of differently abled (human and non) bodies.

By offering a way of thinking via the tangibility and viscerality of visual perception, the flesh reveals the "invisible integuments" (Evernden 1985, 44) that draw bodies together in dynamic relationships. It makes visible and tangible the modes of connection and "intra-action" (Barad 1996) between bodies, both seeing and seen. The flesh works as a bodied metaphor, because it is suggestive of deep visceral sensations that we experience in our own bodies. As such, the flesh is an empathic "tissue." Indeed, Haraway uses the term in this way (though without reference to Merleau-Ponty). For her "the materialized semiosis of flesh always includes the tones of intimacy, of body, of bleeding, of suffering ... one cannot use the word flesh without understanding vulnerability and pain" (2000, 86). Tapping into this bodily (and richly biological) knowledge, Merleau-Ponty's phenomenology can be extended to biological vision to encourage a deeper "feeling for the organism" (Keller 1983) within visualization practices. Extending the metaphor of the flesh, I investigate the possibilities for a biological practice of *seeing with feeling*, one which might work to generate a deeper understanding of the animate lives of nonhuman organisms.

EMBODIED VISION

Haraway's theory of "situated knowledges" (1991) also recognizes the corporeality of vision and draws attention to the embodied nature of all imaging and knowledge practices, especially in science. Our bodies and "prosthetic" technologies form an "optics" through which we vision the world; a complex series of lenses that zoom in and out, drawing our bodies into different relationships with the things that we actively visualize (Haraway 1991). She insists that we must never stop questioning how our bodies and technologies shape the images and meanings we craft. Locating our bodies within and among the multiple and shifting "agencies of observation" (Barad 1996) that inhabit the laboratory offers a means of accounting for our situated, partial positioning and reveals the limitations and contingencies of the knowledge that we can glean from any single vantage point. As such, bearing witness to our implication in visual representations might make us more accountable for the visions we produce of the world.

As a requirement for an embodied ethic of vision, Haraway suggests that "modest witnesses" must account for what they can and cannot see (1997). Accountability for Haraway requires detailed specification of our chosen plane of focus and descriptions of what gets lost as we change position. Simple experiments increasing the magnification of a specimen under a microscope teach us that depth perception and peripheral vision are the first to go, both actually and metaphorically. Situated knowledges advocate taking responsibility for the ways that we make the world visible, and accounting for all the aspects of bodies that our necessarily blinkered vision obscures. To defy the false promises of disembodied objectivity it is necessary to critically situate our participation in visualization practices throughout the many layered processes of knowledge production.

Reminding us that there is no detached or omniscient view, and that our visualizations are power-charged conversations, Haraway asserts that "optics is a politics of positioning" (1991 193). Biological vision is never symmetrical: there is always an element of power wielded by the viewer over the viewed. In a theory that recognizes the materiality of vision, there are consequences for such intimate contact. Yet the violence of biological visualization might lie more in the idealization of vision as some disembodied and "disinterested" system for surveillance. With this assumption of neutrality, biologists are protected from accountability for their actions. The corporeality of vision that Merleau-Ponty and Haraway describe introduces some ethical issues for biological practices when the modalities of vision take on the fleshiness of touch, and visualization technologies acquire the corporeality of bodily prostheses. If their penetrating gaze is reconsidered as a modality of touch, biologists might realize the very uncomfortable intimacy that inheres between themselves and the organisms they manipulate. This might give them pause to reconsider their actions in those crucial moments before they fix or freeze the organism, and render it, at best, as a set of two dimensional marks in their databank of (supposedly) clean, cold, clear observations.

So, can feminist and phenomenological attentiveness to our inter-implication in visual production form a foundation for the integration of embodied ethics into biological practices? Gail Weiss suggests that by recognizing the reversibility of vision and engaging in situated practices like "modest witnessing," we might begin to "see ourselves seeing." Looking at the ways that biology looks at nature, we may also learn to see "how others see our seeing" (Weiss 1999, 43).

BIOLOGICAL IMAGING AND THE DISAPPEARANCE OF TIME, MOVEMENT, AND AGENCY

According to Neil Evernden, Cartesian vision fixes the world and turns it into discrete, bounded objects (1985). Caught up within this paradigm, biologists tend to halt the developmental time of organisms they study in order to draw the internal processes of their bodies into view. They dissect the body's flux and transformation into minute parts, only later to reassemble these static pieces in some logical progression. Such practices produce images of dead bodies that stand-in as life-less representations of what are actually lively processes. I am centrally concerned with the ways that biology tends to disappear the agency of the bodies they draw into view (however partial, decentred, and rhizomous their subjectivities might be). What is this thing called agency, and how can biological imaging practices be used to make it visible or invisible?

Agency suggests intentional engagement in an activity, such as in the active transformation of some thing. It is possible to see the effects of agency at work by interpreting the displacement of things as the "debris" left over in the wake of an action. This is the prevailing method in biology, where dynamic living processes are fixed and frozen in time to make them more visible, quantifiable, and intelligible. But this is not the only way to make agency visible. Another approach would be to attend to the act itself, with an attention to *bodies doing things*.

I am particularly interested in the intimate association between "agency" and the phenomenon of movement. As a biologist and dancer-choreographer, my passion is movement, and I am most interested in the movements of differently bodied beings. While I attend to movement, I do not want to privilege the agencies of "unmarked," able, human bodies. Continually caught-up in dynamic engagements with the world, *all* living bodies (human, nonhuman, and differently abled) are perpetually in motion. Movement is a capacity of all bodies and is manifest in all dimensions of life: from the subtlest gestures in forms of tacit communication between people, to the molecular practices of cells dividing and migrating in the bodies of growing organisms. Drawing attention to the organism-as-subject, we can begin to perceive nonhumans as agents of change both within their own bodies, and within the world.

To explore this relationship between movement and agency further, I draw on Merleau-Ponty's phenomenological investigation of movement as basic "intentionality" (1962). For him, movement underlies all actions as well as all perception and communication, because it is by means of movement that we can aim our bodies at things and take in our surrounding world. He suggests that movement "provides us with a way of access to the world" (1962, 140): we move (sometimes imperceptibly) in order to engage with the world, to express ourselves, and to perceive. Importantly there are many diverse ways to move through, sense, and make sense of the world.

Merleau-Ponty's phenomenology of "movement as agency" can be applied to challenge invasive practices in developmental biology. If movement is one of the possible ways in to perceiving agency, then visualization practices that fix and freeze organisms in order to quantify their life processes inhibit our perception of their capacity for movement, and so erase any indication of their lively activities. "The flesh," that fabric of relationships that extends between all expressive and perceptive bodies (even those who inhabit the laboratory), cannot exist without movement or outside of time. Thus, static imaging strategies in developmental biology (such as scanning electron microscopy, some genetic marker technologies, and in situ hybridization protocols, for example) pose an impediment to the perception of the spatio-temporal dynamics and gestural movements of growth processes. It seems that in its efforts to achieve "body-lessness," the biological visualization has managed to produce a "time-lessness" as well.

Yet, in biology, is it possible to "see without staring" (Evernden 1985), without turning everything under our gaze into inanimate, disembodied, and dismembered objects? Recognizing the full "interpenetration" of gene, organism, and world (Lewontin 2001), some argue that developmental processes must be understood within a wider context that includes both space and time. With the understanding that "form is process," several biologists have revealed the importance of temporality to understanding organismal development (see for example Goodwin 1994, Goodwin and Colquhoun 1997, Ho 1997, Holdrege 1996, Sattler,

1988). The work of these biologists shifts conceptions of development from the determinism of gene-centric models of development, to recognition of the dynamic activities of full-bodied organisms. Merleau-Ponty's philosophy can be applied to such aims to generate a phenomenological description of "the developmental dance" of organisms caught up in processes of "coming into being." Delving into the moving, shape-shifting processes of growth, a phenomenological biologist would be attuned to changes in the whole organism as it expands and extends its body out to meet the world. Such a practice reveals the importance of movement, and therefore temporality, to understanding complexity and self-organization in developmental processes (see Myers 2001). Extending Merleau-Ponty's sense of the intentionality of bodily movements to include the movements of metamorphosis makes visible the lively, embodied nature of developmental processes. His phenomenology complements theories of "contextual biology" (Holdrege 1996) that suggest that it is not genes that make organisms, rather it is organisms who make and remake themselves (see Coen 1999).

If agency can be read through the movements of living organisms (however subtle), then seeing *bodies in time* can make visible the otherwise invisible agencies of metamorphosis. Visualizing the unique temporal dynamics of growing organisms becomes an important prescriptive for technological innovation in biological imaging. Building on this critique of static imaging strategies, I look to the potential contribution of new moving image technologies to theories and practices in developmental biology.

LIVE-CELL IMAGING AND NEW BODILY IMAGINARIES

Visualization technologies that promise the impossibility of 'objective' vision are most often engaged as a means of disappearing scientists' bodies from practices of observation. Yet, can our visualization technologies be used to implicate our bodies, rather than alienate them? Can our bodies' tacit knowledge be brought into play to add depth to biological imaging strategies? Transforming the hierarchical relationships between scientists and nonhuman organisms requires re-membering the embodied nature of vision, re-imagining the perceptive capacities of bodies, and opening up the ontological possibility that nonhuman organisms can exist as more than mere objects of scientific scrutiny.

Haraway challenges feminists to account for the insurgent possibilities of our early-twenty-first-century modes of embodiment as "cyborgs" able to transgress the modernist domains of nature, culture, and technology (1991). In this spirit, I suggest that technologies do exist that might enable biologists to see more fluidly through developmental processes. In a "cyborg-eco-feminist" appropriation of the imaging tools of technoscientific laboratories, I propose that it might be possible to re-embody a range of new visualization technologies and so generate more situated and "time-full" narratives of developmental processes.

There is currently a recognition among developmental biologists that they need better strategies for visualizing the extremely dynamic processes of life. High resolution, time-lapse, and multi-dimensional digital imaging technologies such as laser scanning confocal microscopy are an area of huge expansion in current biological research, and these tools are very rapidly revolutionizing practices in many fields (e.g., Hasselhoff et al. 1999; see also Keller 2002). Time-lapse imaging fills the gaps in-between the freeze frame images of the fixed forms that biologists normally see. "Real-time" imaging allows the observer to move through developmental time with a growing organism. Confocal microscopy enables multi-dimensional imaging of living tissues, and can be used to produce digital videos that record cellular changes over extended periods of time. Sped up, these movies show in intimate detail the movements, involutions, and expansions of developmental growth and physiological activity.[1]

Haraway emphasizes that biological visibility is conditional and limited (1991). How do the limits of technologically mediated vision frame the knowledges we hold of nonhuman organisms? How can changing the limits of vision open up the possibility for new knowledges? These "time-full" moving-image technologies generate extremely rich sets of data. By integrating time, these technologies can generate fluid images that keep pace with the physiological transformations of organisms as they move through space in time. However, the images they produce are contingent, mediated, and strategically framed representations of bodily metamorphoses. Filling in the gaps in bodily visibility is not an unproblematic practice. Haraway cautions that such images are never mere "warped mirror reflections" of the world; they are always diffracted, deflected, and blurred. They have been coded and decoded, digitized and polarized. Learning to read these diffraction patterns is no simple effort of translation; it involves interpreting the trajectories and projections of the light beams of our culture, economy, and politics. The images produced in biological studies are thus the products of complex "mapping practices," (Haraway 1991, 201) where "boundaries" are drawn and "cuts" (Barad 1996, 170) are made between subject-bodies, object-bodies, and an extended range of prosthetic technologies. Distorted by the lenses of culture and ideology, such technologies never provide seamless, value-free images of "nature." They are bold interventions.

Yet, as very new tools for observation, these technologies do give biologists rare glimpses into worlds they might never have imagined. By documenting the qualities, gestures, and details of organismal, cellular, and sub-cellular movements, biologists are "witnessing" the organicist concept of self-organization and the possibility of "intentional" movements in life processes. Seeing *bodies in time* allows for (but does not necessitate) less reductionistic, less mechanistic, modes of visualization and representation in biology. Yet, if these technologies are to reveal organisms as anything more than mere objects, they must be engaged critically with an attentiveness to the limitations, contingencies, and consequences of embodied vision. If engaged with a commitment to situated knowledges, a phenomenological attentiveness to nonhuman agency and movement, and an eco-feminist ethic to account for and respect all bodies involved, these new technologies could generate a new ontology of the organism.

"BODYING" BIOLOGICAL VISUALIZATION TECHNOLOGIES

Understanding how these visual systems work, technically, socially, and psychically ought to be a way of embodying feminist objectivity.
                                                            Haraway 1991, 190.

Haraway suggests that the "elaborate specificity" of an embodied practice of objectivity requires "the loving care people might take to learn how to see faithfully from other's points of view, even when the other is our machine" (1991, 190). Such "loving care" might also be manifest in the development of technologies that can better represent the dynamic, moving lives of those others who are *not* our machines, including those creatures who inhabit our laboratories.

Investigating the relationships between embodiment and "incorporating practices," Katherine Hayles (1999) investigates how technologies transform our bodily practices and thus our modes of experience, representation, and knowing. She suggests that "when changes in incorporating practices take place, they are often linked with new technologies that affect how people use their bodies and experience space and time" (1999, 205). Each technology reconfigures relationships between biologists and organisms, generating new and different kinds of knowledges. "Incorporating" these new biological imaging technologies will definitely impact the ways that organisms and their developmental processes are conceived, represented, and interpreted both scientifically and culturally. However, these technologies must be engaged critically and cautiously. Feminists, environmentalists, and citizens more broadly need to understand how biological imaging technologies work to shape our cultural representations and expectations of the living world.

An embodied approach to ethics might be the best way to keep pace with our shifting relationships and responsibilities as we integrate these evolving technologies into our practices. Embodied ethics may help to establish more accountable frameworks for the kinds of relationships that biologists structure with laboratory organisms. Applying these technologies towards projects "less aligned along axes of domination" (Haraway 1991, 192), I envision moving-image technologies used to generate an ontological shift to transform the ideologies, questions and practices that orient developmental biology research. A *cyborg-eco-feminist* project might investigate ways of using moving-image technology as an educational and research tool to transform the prevailing narratives of nature in developmental biology. Moving beyond mechanistic, neo-Darwinian tales of evolution and development, these new imaging technologies can be used to extend the "biological imaginary" of Western culture towards narrating richer, more "body-full" representations of the world.

NOTES

1 Given the constraints of the printed page, live-cell images are not reproducible here. There are several web sites that do have live-cell movie clips available for downloading. One excellent site is Jim Hasselhoff's (University of Cambridge, UK) laboratory web site at: http://www.plantsci.cam.ac.uk/Haseloff/Index MOVs.html (accessed November 2003). Given the temporal, filmic nature of these images, the media they move through most readily are the Internet, CD-ROMs, and digital video. Not reproducible in the traditional format for scientific publication in print journals, live-cell images find other homes. Increasingly, scientific journals are using online formats to link publications with their supporting live-cell and confocal data, as well as other add-on features accessible only on the Web. As mesmerizing images, their aesthetics are put to work in cinematic form to entice and entertain audiences at science centres and in school classrooms. In live-cell movie clips, cells "perform" for their audiences, adding dynamism to power point presentations at scientific conferences and seminars, and bringing experimental results to life.

And yet, with such aesthetic appeal, live-cell images are also enlisted to do "extra-scientific" work as public relations tools on the Internet. They are found most readily on laboratory web sites, promoting a given scientist's research program; as links embedded in online cell imaging "galleries" and interactive teaching tools which feature "cool" bioimaging sites and downloadable movie files; as publicity for new bioimaging centres; and as advertisements on manufacturers web sites, selling live-cell microscopes and their accoutrements.

REFERENCES

Amann, K., and Knorr-Cetina, K. 1990. The fixation of (visual) evidence. In *Representation in scientific practice*, eds. M. Lynch and S. Woolgar, 85–122. Cambridge, MA: MIT Press.

Barad, K. 1996. Meeting the universe halfway: Realism and social constructivism without contradiction. In *Feminism, science and the philosophy of science*, eds. Nelson, L.H., and Nelson, J., 161–94. Boston: Kluwer Academic Publishers.

Coen, E. 1999. *The art of genes: How organisms make themselves*. Oxford: Oxford University Press.

Daston, L. and Galison, P. 1992. The image of objectivity. *Representations* 40, 81–128.

Evernden, N. 1985. *The natural alien*. Toronto: University of Toronto Press.

Galison, P. 1998. Judgement against objectivity. In *Picturing science, producing art*, eds. A. Jones and P. Galison, 327–59. New York: Routledge.

Goodwin, B. 1994. *How the leopard changed its spots*. London: Weidenfield and Nicolson.

Goodwin, B. and Colquhoun, M. 1997. In context: Genes, organisms and evolution illustrated through algae and buttercups. In *The future of DNA*, eds. J. Wirz and E.L. van Bueren, 181–97. Dordrecht: Kluwer Academic Publishers.

Haraway, D. 1991. *Simians, cyborgs and women: The reinvention of nature*. New York: Routledge.

– 1997. *Modest_witness@second_millennium.femaleman©_meets oncomouse™*. New York: Routledge.

– 2000. *How like a leaf: An interview with Thyrza Nichols Goodeve*. New York: Routledge.

– 2001, February. *From cyborgs to companion species: Kinship in technoscience*. Keynote address presented at the conference Taking Nature Seriously: Citizens, Science and Environment, University of Oregon, Eugene. February 25–27, 2001.

Hasselhoff, J. et al. 1999. Live imaging with green fluorescent protein. In *Methods in molecular biology: Confocal microscopy methods and protocols*, ed. S. Paddock, 241–60. Totawa, N.J.: Humana Press.

Hayles, N.K. 1999. *How we became posthuman: Virtual bodies in cybernetics, literature and informatics*. Chicago: University of Chicago Press.

Ho, M-W. 1997. DNA and the new organicism. In *The future of DNA*, eds. J. Wirz and E.L.van Bueren, 78–93. Dordrecht: Kluwer Academic Publishers.

Holdrege, C. 1996. *A question of genes: Understanding life in context*. Hudson, New York: Floris Books.

Keller, E.F. 1983. *A feeling for the organism: The life and work of Barbara McClintock*. New York: W.H. Freeman and Co.

– 1996. The biological gaze. In *Future natural: Nature, science, culture*, eds. G. Robertson, et. al., 107–21. New York: Routledge.

– 2002. *Making sense of life: Explaining biological development with models, metaphors and machines*. Cambridge, MA: Harvard University Press.

Latour, B. (1990). "Drawing things together." In *Representation in scientific practice*, eds. M. Lynch and S. Woolgar, 19–68. Cambridge, MA: MIT Press.

Lewontin, R. 2001, February. *The interpenetration of gene, organism and environment*. Keynote address presented at the conference Taking nature seriously: Citizens, science and environment, University of Oregon, Eugene. February 25–27, 2001.

Lynch, M. 1990. The externalized retina: Selection and mathematization in the visual documentation of objects in the life sciences. In *Representation in scientific practice*, eds. M. Lynch and S. Woolgar, 19–68. Cambridge, MA: MIT Press.

Merleau-Ponty, M. 1962. *The phenomenology of perception*. (C. Smith, trans.). New York: Routledge.

– 1968. *The visible and the invisible*. (A. Lingis, trans.) Evanston:Northwestern University Press.

Myers, N. 2001. *"Body-fullness" in biology: Feminist environmentalism meets Merleau-Ponty's philosophy*. MA thesis, Faculty of Environmental Studies, York University, Canada.

Sattler, R. 1988. A dynamic multidimensional approach to floral morphology. In *Aspects of floral development*, eds. P. Leins, S.C. Tucker and P. K. Endress, 1–6. Berlin: Stuttgart.

Weiss, G. 1999. *Body images: Embodiment and intercorporeity*. New York: Routledge.

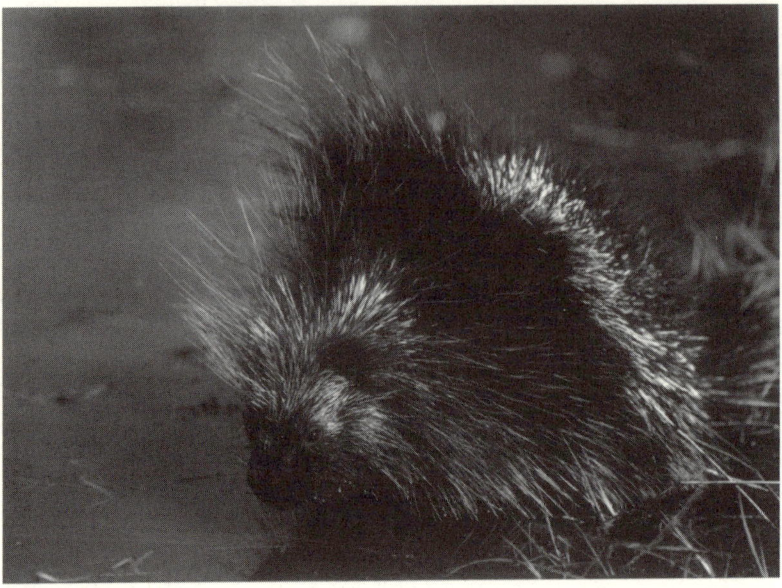

18

Bioregional Teaching: How to Climb, Eat, Fall, and Learn from Porcupines

LEESA FAWCETT

IMAGINING PORCUPINES

No one ever taught me about porcupines in school and I wish they had. At dusk, a solitary porcupine and I would often pass each other in our neighbourhood. As a puppy, my dog Ruby ran into a few porcupines (literally) until she learned to give them their space. The porcupine (*Erethizon dorsatum*) is the only mammal in North America that is covered with long sharp quills (which, incidentally, they cannot throw at humans). They are short-legged vegetarians who are more comfortable in trees than on the ground. The *Peterson North American Field Guide to Mammals* (Burt and Grossenheider 1976) actually calls them "clumsy" animals. There is fossil evidence of porcupines from the Oligocene epoch – more than 36 million years ago; so clumsy or not, they have successfully inhabited this earth for a very long time.

I learned, surprisingly, that porcupines often have broken bones. They climb high up in trees and out onto thinner branches to feed, and sometimes they fall. Given their thickly covered quilled bodies this seems like an odd dilemma, as they must impale themselves on their own sharp spines. Porcupine researcher Uldis Roze made a fascinating discovery after a porcupine struck him with a powerful tail slap, thereby lodging quills in his upper arm. Roze (1989) pulled out the quills he saw but there was a painful bulge where one quill had entered his arm. Several days later, when he was teaching, he felt a sharp quill catch the inside of his shirt and ten inches below the original point of entry there was a red mark, and suddenly his pain dissipated. Roze wrote: "The quill had emerged! I flexed and twisted the arm luxuriously. I found the quill inside my jacket arm ... It looked beautifully clean, scoured by the body to a glistening freshness. Like some mole of the flesh, it had ratcheted its way past muscles, nerves, and blood vessels to emerge far from the site of entry" (1989, 24).

Roze collaborated with biochemists and discovered that porcupine quills are coated with antibiotic chemicals (Roze et al. 1990). In fact, eight percent of a porcupine quill's weight is a form of grease that contains fatty acids with antibiotic properties (Roze 1989, 27). A porcupine can fall out of a tree, be struck by its own quills, and those quills can pass through its body and out again, often with very little harm done. I think porcupines serve as an intriguing metaphor for the ability to climb high, nourish oneself, rest, and sometimes fall, break bones, and still be protected against one's own defences and heal. Learning and teaching environmentally could benefit from an understanding of this process of climbing, slipping, and falling to earth – learning and healing from our prickly mistakes. Roze (1989) who has spent over a decade closely studying porcupines wisely noted that, "Again and again, the porcupine has been a teacher, a storyteller of the woods, a complexifier and adorner of the world" (15). What better purpose is there for teaching and learning than to complexify and adorn the world?

Porcupines are neither globally distributed nor homogeneously similar. They are living subjects who cannot survive anywhere and eat anything. Actually, Roze (1989) found that individual porcupines had innate preferences for certain tree species; one fed consistently on hemlock, another on beech, and yet another on sugar maple. (Plants can secrete their own metabolic defences against these predators to affect the relationship.) In Western societies, porcupines are not considered useful; in fact many people treat them as pests and, consequently, humans are their chief predators. This is in stark contrast to some Indigenous peoples, who for centuries employed and honoured porcupines for food, crafts, and art, such as fine quillwork (Roze 1989). Today, porcupines are not a valuable resource and they are not on any school curriculum that I know of. What would it mean to learn about porcupines? It might mean that students would gain local intimate knowledge of another life. It might mean that we would slow down in time and space, and quietly pay attention to something other than ourselves. It might mean that teachers and students would go out into the field to carefully observe and use all their senses to witness moments in the life of another species. If current science and environmental education are critiqued for being homogeneous and disengaged from daily life, then bioregional knowledge offers an antidote. If education in general is criticized for being anthropocentric, then learning about porcupines can free us from the constraints and oppressions of a made-for-humans-only world. Good quality environmental education aims to be interdisciplinary, participatory, critical, community-based, and values-based (Hart et al. 1999). I am left with a persistent belief that someone should have taught me about porcupines (or any other local species such as chickadees, toads, garter snakes, raccoons, flies, etc.). To do this, to "open up our experience (and, yes, our curricula) to existential possibilities of multiple kinds is to extend and deepen what each of us thinks of when he or she speaks of a community" (Greene 1995, 161).

NATURAL HISTORY

In science and environmental education, and in our culture generally, we do not pay adequate attention to increasing people's awareness and knowledge of local, common animals (Nabham and St. Antoine 1993). If children, for instance, were encouraged to experience animals in their bioregion in sensory and intentional ways, then their curiosity, fears, and passions could emerge and significantly shape science curricula. I believe the results would be much more rewarding for the children, and ultimately beneficial for the conservation of multi-species communities. I side with feminist philosophers Maxine Greene and Hannah Arendt who "believed that nothing can keep a community together when people lose their interest in a common world" (Greene 1995, 196). For me, the catch is that communities must be inclusive of all the species present, not just the humanoids. Arendt (1961) said that "Education is the point at which we decide whether we love the world enough to assume responsibility for it" (in Greene 1995, 196). In order to love the world well we need to know it intimately, and learning local natural history is one way to act on this affection.

There has been a recent call for a revival of natural history knowledge, as it appears to have become endangered. In his research on multiple intelligences, Gardner (1999) has newly recognized "naturalist intelligence" – expertise in the recognition and classification of local flora and fauna – as a distinct invaluable form of learning. Specific pattern recognition is a vital skill for any person to hone, and the elaboration and comparison of patterns is foundational to discovery processes in science (Loehle 1994), and to the talents of poets, artists, farmers, and cooks (Gardner 1999). These "naturalist" qualities of curiosity and attentiveness to life patterns should be actively encouraged in all citizens. In the sciences, E.O. Wilson (2000) argues, biological conservation and community ecology desperately need students of natural history.

Ecology was supposed to be "the subversive science," but I agree with Michael Quinn who wondered if "subversion was now the proper task of natural history" (1995, 8). Natural history can be an unruly science where unexpected things emerge. Amateur naturalists who study local life forms for years on end can learn about long-term changes to populations, habitats, and ecosystems. Often naturalists have other professions and specialties and the cross-fertilization of knowledge can spur discoveries. That kind of intimate, longitudinal, interdisciplinary thinking and data is hard to come by, and is exactly what is needed to comprehend and address patterns of biological deterioration. The story of the Peppered Moth (*Biston betularia*) is a case in point.

The research on the Peppered Moth traces the process of colour change from light to dark in certain moth populations following the Industrial Revolution in England (mid-eighteenth century). Industrialization brought with it a drastic increase in air pollution (particularly emissions of sulphur dioxide), and subsequently an increase in the spread of black forms of the peppered moth. Industrial melanism is evident when the distribution of blacker moth forms is in, and around, major sources of industrial pollution, and displaced in the direction of the prevailing winds. The standard studies of industrial melanism were pioneered by H.B.D Kettlewell from 1955 to 1973 and summarized in his classic book, *The Evolution of Melanism*.

At the age of forty-five, Kettlewell, a medical practitioner and enthusiastic lepidopterist, began a Nuffield fellowship to study industrial melanism. In 1956, after a national medical survey found a disproportionate number of bronchitis victims in certain rural areas of England, Kettlewell was asked to use melanic moth frequencies as indicators of pollution levels. A full discussion of industrial melanism and natural selection is beyond the scope of this paper, but Kettlewell did believe that the darker forms of the moth had a cryptic advantage on the soot blackened trees of England and were preyed upon less than the conspicuous lighter forms. What does matter a great deal is the role amateur naturalists played in unravelling the relationships between air pollution, moth populations, and human and ecosystem health. Kettlewell (1958) recruited 150 naturalist observers who gathered over 20,000 records of the peppered moth and its melanic forms. Many of these observers were keen backyard natural historians who had been keeping records of local species for decades.

The "Peppered Moth Story" would have had a cohesive, happily-ever-after-ending if we could claim that the Clean Air Act passed in Britain was in response to the documented work of naturalists and scientists. Instead, the cancellation of an opera performance because the audience could not see the stage, and the death of prize cattle at a national show were the catalysts for the 1956 Clean Air Act (Berry 1990). A fickle but eventually laudatory response to smog was born. This is similar to falling from a tree and not necessarily learning much from it.

ECOLOGICAL FEMINISM AND BIOCENTRIC ETHICS

Natural history skills and knowledge will not be enough. I concur with Heshusius (1994) who states that educational researchers should describe their work in ethical and participatory terms, not just methodological ones. Natural history knowledge integrated with a biocentric ethics is needed. I purposefully use the term *biocentric ethics* as opposed to *environmental ethics* because I fear that the word *environment* is losing meaning. Environment has become one of those Uwe Porksen plastic words "that has been used in science, given authority by virtue of its use there, and then relocated back into the vernacular, where it sounds important but doesn't really mean anything." (Evernden 1992, 115). Joy Williams, in *Ill Nature*, writes expressively about how tiring the word environment has become.

Such a bloodless word. A flat-footed word with a shrunken heart. A word increasingly disengaged from its association with the natural world. Urban planners, industrialists, economists, developers use it. It's a lost word, really. A cold word, mechanistic, suited strangely to the coldness generally felt towards Nature. It's their word now. You don't mind giving it up. (2001, 5)

It is time to return to the living and leave the cold, deadened words like *environment* behind for a while. *Biocentric ethics* is centred on the bios and all of its members; it signifies the living and their needs, desires, and gifts to one another. Many ecological feminist scholars have analyzed the multiple oppressions acting in Western culture (racism, sexism, homophobia, classism, able-bodied bias), and have enlarged our knowledge by including "speciesism" as one of the interlocking oppressions that hold larger systems in place (Haraway 1991; Plumwood 1997). As science fiction writer Ursula LeGuin writes:

By climbing up into his head and shutting out every voice but his own, "Civilized Man" has gone deaf ... He hears only his own words making up the world. He can't hear the animals, they have nothing to say. Children babble, and have to be taught how to climb up into their heads and shut the doors of perception. No use teaching women at all, they talk all the time, of course, but never say anything. This is the myth of Civilization, embodied in the monotheisms, which assign soul to Man alone (1987, 11).

Focusing on ecological feminism as a critical, socially-minded environmentalism set on generating anti-oppressive moral insights, Chris Cuomo (1998) calls for an ethic of flourishing – an ethical commitment to the well-being or flourishing of communities, species, and individuals. Orr (2001) suggests science should belong to a larger moral ecology that is biocentric in purpose. In recent research, I found that many children expressed friendship, empathy, and compassion for three common wild animals (Fawcett 2002). These feelings gesture towards the foundations of an inter-species morality, and, consequently, teaching should honour and build on children's desires and abilities. Empathetic knowledge can nourish discovery and certainly contributes to a child's interest in animals.

Cheney and Weston (1999) persuasively outline the need for an environmental ethics-based epistemology, in contrast to the reigning epistemology-based ethics. In the former, ethical action is predominately a way to enrich the world and create more possibilities for deeper knowing; in the latter, ethical action is in strict response to knowledge of nature. To face a porcupine as a subject unto itself, to acknowledge that intersubjective space, is to try and stand ethically before the animal. Cheney and Weston (1999) advocate an ethics-based epistemology on the grounds that: a) the world is neither easily nor simply knowable; b) ethics is not extensionist and incremental, but pluralistic and dissonant; and c) because hidden possibilities surround us the task of ethics is to call them out to illuminate and improve the world.

Useful distinctions among nature-as-object, nature-as-self, and nature-as-miracle are made by Evernden (1988, 1992). Nature-as-object is the dominant and familiar belief that nature is a storehouse of resources, a "bare-bones nature with no subjectivity and no personal variables at all: just stuff" (Evernden 1988, 11) – stuff, such as pesky dandelions to be dealt with, and commodities, such as oil, gold, water, and wheat to buy and sell.

In this three-dimensional world of dualisms, nature-as-self arises when we lose the idea "that we are merely skin-encapsulated egos, and realize that we actually have a 'field of care' in which we dwell, which makes us literal participants in the existence of all beings, [only] then we will realize that to harm nature is to harm ourselves" (Evernden 1988, 12). Whilst Nature as extended self has been the platform of the deep ecology movement, ecological feminists have critiqued this form of identification for its emphasis on personal transformation at the expense of political and social structures and inequities (Kheel 1990, Warren 1990, Plumwood 1993). Plumwood (1993) has warned that the resolution of dualism does not require merger or the simple erasure of the boundary between the colonizer and the colonized. Similarly, Evernden (1992) has questioned how different these two positions really are: Nature-as-object begs the question, What's in it for me? While nature-as-self asks, What is this to me?

Nature-as-miracle refers to the wondrous, the inexplicable and unpredictable, and asks the metaphysical question, What is this? (Evernden 1988) It is not meant to be miraculous in an exclusively religious, hand-of-God-sense. A scientist thinking about miracles, Eiseley wrote:

Since ... the laws of nature have a way of being altered from one generation of scientists to the next, a little taste for the miraculous in this broad sense will do us no harm. We forget that nature itself is one vast miracle transcending the reality of night and nothingness. We forget that each one of us in his [sic] personal life repeats that miracle (1978, 291).

One of the emergent possibilities in human relationships is to be in awe, to wonder, to be enchanted by encounters and experiences with other life. Certainly Indigenous cultures have had respectful and integral relationships with the surrounding plant and animal life, as the Haudenosaunee Creation Story and Thanksgiving Address illustrate (Haudenosaunee Environmental Task Force 1999). Beyond awe and empathy lies the terrain of participatory consciousness – forms of sensory, somatic knowing, embodied knowing that does not dissolve into nature just like-me, or nature as transcendent force. A crucial question for socially and environmentally just people is: What does a participatory consciousness of difference and kinship have to offer biocentric ethics and practices?

Ehrenfeld (2000) writes that huge forces such as economic globalization sweep away local biodiversity, local memory, and local autonomy. I believe, along with many of my colleagues, that the depletion of biodiversity is an intensely value-laden topic that runs counter to constitutional democracy. Wood (2000) argues that biodiversity is an intrinsic public good and that democratic policy must be reformed to consider long-term public preferences. With her knowledge of radical democratic politics, eco-feminist Sandilands (1999) invigorates debates about how to include "nature" in democratic conversations; she emphasizes the sociality and emancipatory potential of our animalness, using the common metaphor of "political animal" to signify political and ethical possibilities. American philosopher Hargrove (2000) discusses Environment Canada's concept of environmental citizenship and reminds people never to sever environmental ethics from citizenship. In the skin of our "political animalness" what kind of communities do we imagine and make?

Western culture teaches children to divorce themselves from their "animalness." Yet I believe that the opportunity to experience various species, and to differentiate between other animals is important to a child's sense of self, and to a child's sense of itself as a human being, beyond being an individual. Greene (1995) perceptively writes that: "Our very realization that the individual does not precede community may summon up images of relation, of the networks of concern in which we teachers still do our work and, as we do so, create and recreate ourselves" (197). Nelson (1993) shifts the emphasis from individual knowledge-making to epistemological communities – or communities of knowers. I contend that the community of knowers and what is to be known is a multispecies community – an "ecology of subjects." The relationships of power are of course unequal in these communities, which is why we need something like biocentric ethics, so we can build towards far-reaching environmental citizenship.

BIOREGIONAL FIELD COURSE

To ground ideas about a political "ecology of subjects" and environmental citizenship in praxis, I will discuss a bioregional field course my colleague Robert MacDonald (a physicist who works on strategies for sustainable development and energy) and I undertook in the spring of 2001. I do not ascribe to one fixed definition of bioregionalism, but a bioregion is generally described as a cultural geographical area defined by common natural features such as watersheds, landforms, local ecosystems and biotic shift, indicator species, and human history. Jim Dodge (1990) writes that "bioregionalism is an idea still in loose and amorphous formulation" (5) that has to do with resistance, renewal, and reinhabitation, and can be characterized by "a decentralized, self-determined mode of social organization; a culture predicated upon biological integrities and acting in respectful accord; and a society which honours and abets the spiritual development of its members" (10).

In the Faculty of Environmental Studies at York University, where Rob and I work, teaching and learning environmentally signifies transdisciplinary ways of thinking about natural, social, built, and organizational environments. We envisioned this field course as a form of environmentally based community pedagogy, focusing on the Niagara Escarpment and Headwaters area of Dufferin County, Ontario. Headwaters County is so named because four significant rivers – the Grand, Credit, Humber, and Nottawasaga – originate in its hills. In addition, two significant Ontario geographical features come together to create rare and beautiful landforms – the Niagara Escarpment, which has been recognized as a World Biosphere site, and extensive glacial moraines. Volunteers maintain the renowned Bruce Trail, which offers public access to this vital region.

Dufferin County is situated immediately north of the Greater Toronto Area and is experiencing increasing pressure for economic development. The field course was designed for fourth-year undergraduate and first-year graduate students from a variety of backgrounds to work together in a field setting with local citizens on pressing bioregional challenges. We chose to work with the Mono-Mulmur Citizen's coalition to contribute to the development and implementation of a sustainable countryside vision, which preserves, restores, and enriches the ecological and cultural heritage of Headwaters County.

Cheney (1989) urges communities to tell the best stories they can about living in particular places, as a way to develop more accountable bioregional environmental ethics. Following Cheney, we organized a list of diverse local speakers to tell us their stories about this bioregion including representatives from: Mono-Mulmur Citizen's Coalition, Dufferin County Museum and Archives, Land Stewardship Network, Nottawasaga Conservation Authority, Orangeville Farmer's Market, Headwaters County Tourism Association, Organic farmers, Hereford Beef farm, Mono Township planning, Orangeville city planning, Niagara Escarpment Commission and Caledon Countryside Alliance.

Direct experience of nature; inclusive, diverse communities; and respect for trial and error are necessary attributes for community-based valuing to work, according to eco-feminist Lori Gruen (1997). This field course gave participants opportunities to directly experience the bioregion, as we lived at a local ecologically run retreat centre and went on many hikes and walkabouts in the area. At the end of the course, at a public meeting, students presented their varying analyses and suggestions back to the community for discussion. So we witnessed the pleasurable and untidy gathering of diverse communities of knowers thinking about bioregional citizenship (or as my son succinctly says "It was Cool!").

In his current book, *The Enemy of Nature*, Kovel (2002) writes unwaveringly about the breadth and depth of capitalism's ecological and social destruction. Unfortunately, he critiques bioregionalism and its notions of self-sufficiency as too essentialist (Kovel 2002). This simplistic critique suffers from not realizing the moments and forces of strategic essentialisms in collective movements for social change. There are always advantages, disadvantages, and unpredictable outcomes at play when people cooperatively identify with any group, but out of those spaces, practices and coalitions can form that can transform dominant relations. In her in-depth discussion about essentialisms in practice, Noel Sturgeon writes: "Essentialist rhetoric and theorizations within oppositional social movements should be recognized as complex deployments within particular social, political, and historical contexts with ambivalent, contradictory outcomes" (1997, 11). Bioregionalism is one example of a complex articulation and lived practice.

I do not believe that bioregionalism is the be-all-and-end-all, but I do think it is an environmentally and socially conscious movement that redirects our attention back to green local places and people. In this age of relentless globalization, this is not a bad oppositional direction with promising counter-hegemonic possibilities. As Roze (1989) wrote: "A study of the porcupine requires, above everything, time" (ix). Time to watch, listen, and be attentive to another life force in all of its complexity and mystery. Curiosity about life and the agency of other beings is one antidote to the hyper-real world of nanosecond-long messages and popular cultures that suffer from attention deficit disorders. Watching porcupines in their neighbouring and interdependent life-world may help more than we can currently imagine.

REFERENCES

Berry, R. 1990. Industrial melanism and peppered moths. *Biological Journal of the Linnean Society*, 39, 301–22.

Burt, W. and R. Grossenheider. 1976. A field guide to the mammals. (Peterson Field Guide Series), Boston: Houghton Mifflin Company.

Cheney, J. 1989. Postmodern environmental ethics: Ethics as bioregional narrative. *Environmental Ethics* 11(2), 117–34.

Cheney, J. and W. Weston. 1999. Environmental ethics as environmental etiquette: Toward an ethics-based epistemology. *Environmental Ethics* 21, 115–34.

Cuomo, C. 1998. *Feminism and ecological communities: An ethic of flourishing.* New York: Routledge.

Dodge, J. 1990. Living by life: Some bioregional theory and practice. In *Home! A bioregional reader*. eds. Van Andruss, C. Plant, J. and E. Wright. Gabriola Island, BC: New Society Publishers.

Ehrenfeld, D. 2000. War and peace and conservation biology. *Conservation Biology* 14(1): 105–112.

Eiseley, 1978. *The star thrower.* New York: HBJ publishers.

Evernden, N. 1988. Nature in industrial society. In *cultural politics in contemporary America*, eds. S. Jhally and I. Angus, New York: Routledge, Chapman and Hall.

– 1992. *The social creation of nature.* Baltimore: The John Hopkins University Press.

Fawcett, L. 2002. Biological conservation of common and familiar animals: The roles of experience, age and gender in children's attitudes towards bats, frogs and raccoons. PhD dissertation, York University.

Gardner, H. 1999. *Intelligence reframed: Multiple intelligences for the 21st century.* New York: Basic Books.

Greene, M. 1995. *Releasing the imagination: Essays on education, the arts, and social change.* San Francisco: Jossey-Bass Publishers.

Gruen, L. 1997. Revaluing nature. In *Ecofeminism: Women, Culture, Nature,* ed. K. Warren, 356–74. Bloomington: Indiana University Press.

Haraway, D. 1991. *Simians, cyborgs and women: The reinvention of nature.* New York: Routledge.

Hargrove, E. 2000. Toward teaching environmental ethics. *Canadian Journal of Environmental Education* 5, 114–33.

Hart, P., Jickling, B. and R. Kool. 1999. Starting points: Questions of quality in environmental education. *Canadian Journal of Environmental Education* 4, 104–124.

Haudenosaunee Environmental Task Force. 1999. *Words that come before all else: Environmental philosophy of the Haudenosaunee.* Cornwall Island, Ontario: Native North American Travelling College.

Heshusius, L. 1994. Freeing ourselves from objectivity: Managing subjectivity or turning toward a participatory mode of consciousness? *Educational Researcher* 23(3), 15–22.

Kettlewell, H. 1958. A survey of the frequencies of *Biston betularia* (L.) (Lepidoptera) and its melanic forms in Great Britain. *Heredity*, 12, 51–72.

– 1973. The evolution of melanism. Oxford: Clarendon Press.

Kheel, M. 1990. Ecofeminism and deep ecology. In *Reweaving the world,* eds. I. Diamond and G. Orensten, 128–37. San Franscisco: Sierra Club Books.

Kovel, J. 2002. *The enemy of nature: The end of capitalism or the end of the world?* Halifax, Nova Scotia: Fernwood Publishing Ltd.

LeGuin, U. 1987. *Buffalo gals and other animal presences.* New York: New American Library.

Loehle, C. 1994. Discovery as a process. *The Journal of Creative Behaviour* 28(4), 239–50.

Nabham, G. and S. St. Antoine. 1993. The loss of floral and faunal story: The extinction of experience. In *The Biophilia Hypothesis*, eds. S. Kellert and E. Wilson, 229–250. Washington, DC: Island Press.

Nelson, L. 1993. Epistemological communities. In *Feminist Epistemologies*, eds. Alcoff, L. and E. Potter. New York: Routledge.

Orr. D. 2001. A literature of redemption. *Conservation Biology* 15(2), 305–7.

Plumwood, V. 1993. *Feminism and the mastery of nature.* New York: Routledge.

– 1997. Androcentrism and anthropocentrism: Parallels and politics. In *Ecofeminism: Women, culture, nature*, ed. K. Warren, 327–55. Bloomington: Indiana University Press.

Quinn, M. 1995. *Natural history and Ontario birders: The natural history of a collector.* PhD dissertation. Toronto, York University.

Roze, U. 1989. *The North American porcupine.* Washington, DC: Smithsonian Institution Press.

Roze, U., Locke, D. and N. Vatakis. 1990. Antibiotic properties of porcupine quills. *J. Chem. Ecology* 16.

Sandilands, C. 1999. *The good natured feminist: Ecofeminism and the quest for democracy.* Minneapolis: University of Minnesota Press.

Sturgeon, N. 1997. *Ecofeminist natures: Race, gender, feminist theory, and political action.* New York: Routledge.

Warren, K. 1990. The power and promise of ecological feminism. *Environmental Ethics* 12(2), 121–46.

Williams, J. 2001. *Ill nature: Rants and reflections on humanity and other animals.* New York: Vintage Books.

Wilson, E.O. 2000. On the future of conservation biology. *Conservation Biology* 14(1), 1–3.

Wood, P. 2000. *Biodiversity and democracy: Rethinking society and nature.* Vancouver: UBC Press.

Index

Aboriginal peoples. *See* Indigenous peoples, knowledges
academia/academic, 50, 53, 57, 61, 108, 112, 127, 134, 149–61, 181–91, 196–207, 211, 226
 – academic capitalism, 149–61
 – architecture, 155, 161
 – "chilly climate," 134, 154, 196–201. *See also* women; higher education; equity
 – disciplines, 225. *See also* Eurocentrism
 – discreditation/centring: of anti-colonial thought, 4; of caring discourses, 160, 198, 210; of environmental issues, 183, 187, 189, 197; of feminist science, 1–2, 70–2, 108, 152, 160; of Indigenous knowledges; *see* Indigenous knowledges
 – and environment, 153, 182–7, 189, 197–8
 – and equity. *See* equity
 – evaluation. *See* evaluation
 – funding. *See* research; retrenchment
 – gendered and racialized, 158
 – intellectual property. *See* capitalism
 – interdisciplinarity, 202, 229
 – limits of academic discussion of spirituality, 53, 57
 – and 'logic of accounting,' 155
 – research. *See* research
 – retrenchment, 155, and escalation of inequity, 158–9
 – sociology, 189
 – and teaching. *See* teaching
 – women in. *See* women, professional education
activism, bioregionalism as, 278; and centring spirituality, 56–7, 61; and decolonization, *see* decolonization; environmental, 150, 182; and equity at the university, 159, *see also* academia; equity; feminist, xiii–xv, 7, *see also* knowledges; science, feminist; about genetically modified foods, 171; and hope, 75, 77, 191, *see also* environmental sustainability, teaching; by Indigenous peoples, 56, 61, 126, 152, 220, 227–31, 236, *see also* colonization, anti-colonialism; and need for literacy about science, 181–92, 265, *see also* languages/literacy; for peace, xiii–xiv, 61, 151–2, 185–6, *see also* nuclear issues; public awareness of science and passive resistance, *see* public; and research, 15, 166; rupturing hegemonic practices, 49–50, 56–9, 74, 142–3, 152, 227–8, 278; and strategic essentialism, 278; around sustainable science, 67, 71, 166, 198; teaching, *see* teaching, as activism
agency. *See* scientific method, activism
anthropology. *See* colonization

anti-colonialism, 56, 58. *See also* colonization
anti-racism. *See* racism
architecture, on university campuses, 155. *See also* professions
Asking Different Questions, 6, 66, 90–2, 161, 198;

biology, biologists. *See* science, environmental sustainability, Nature
Biology As If the World Mattered (BAIT-WORM), 5, 6, 8, 166
biopiracy. *See* biotechnology, colonization
bioregionalism, definition, 276. *See also* activism
biotechnology/technology, 14–15, 56, 85–7, 123, 151–2, 155–6, 167–76, 182–6, 206, 255; biopiracy and ethnoscience, 111, 123, 151, 184, 226–7, *see* also colonization, biopiracy; racism, global; critique/dangers/consequences of, 14, 48–50, 85–7, 123, 151–2, 168–72, 185–6, 192n; and democracy, 14, 168, 171, 182; and the domain of medical laboratory technology, 85; feminist Indigenous vision, 48, 54–6; genetically modified organisms, 152, 155, 166–73, 184, 185–6, 192n; acceptance/assessment of risk, 168–70, 186; vs insurgent technology, 16; and the precautionary principle, 186; and promises for a sustainable future, 171–2; and public distrust, *see* public; research funding for, 172; social impact of, taught in engineering, 203; state support for (neoliberalism), 171–3; in university curricula, 156, 159, 182–3; user-friendly, 206

capitalism, 13, 55, 65–7, 70, 122–3, 135–42, 150–7, 170–3, 176, 188, 198, 227, 278; academic, *see* academic, capitalism; and colonization, *see* colonization; and the corporate elite, 136–7, 188; critique of, *see* critique; and the direction of agriculture, 172–3; ethnoscience and multinational corporations/biopiracy, 55, 123, 137, 151, 161, ; 169, 181, 226–7, *see also* colonization, Indigenous knowledges; exploiting Third World, 137, 136–7, 188; and forestry industry/research/professional education, 65–7, 123,150; and intellectual property, 123, 137, 161, 176; and pedagogy, *see* pedagogy; pharmaceutical industry/research, 14, 15, 123,149, 155–6, 169, 171, 185, 226–7, *see also* biotechnology; and the privatization of public goods, 123, 173; and professional education/origins of professions, 153–5; as sacred, 137–8; transnational corporations, 135. *See also* academic capitalism; development; globalization, and dehumanizing effects of restructuring; Third World
caring. *See* curriculum, caring; evaluation
"chilly climate." *See* academia
class: and professions, 153–4. *See also* oppressions
colonization/neocolonialism, 50, 123–30, 134–9,151, 224–6 ; agricultural, 5, 151, 156, 171–2 ("molecular conquistadors"); and anthropology/linguistics, 122, 130; anti-colonial, discursive framework, 50, 58–9, cause, 56; and biopiracy, *see* capitalism; and capitalism, 14, 55, 123, 134–43, 151, 172; consciousness-raising, 9–12, 58; critique of biotechnology, *see* biotechnology; critique of Eurocentrism, 13, 54–6, 58, 113, 122–8, 139, 167, 225; critique of science and pedagogy, 17, *see also* critique; globalization as, 14, 48, 55, 115, 121, 132–45, 151, 167; imperialism, 50, 121–4, 126 (cultural), 134, 225–6 (cognitive); of India, 134; of Indigenous knowledges (appropriation/biopiracy), 55–6, 73, 113–14, 121–5, 156; and the Internet, 55–6; and maps, 29; relationship with capitalism, 14, 55, 123, 135–8, 151, *see also* biotechnology; role of education, 127, 224; role of White women scientists, 113–15. *See also* decolonization; post-colonialism
community/ies, 271, 274, 276; epistemological, 276. *See also* Indigenous peoples
consciousness-raising. *See* decolonization; pedagogy; spirituality
critique, 1, 13–14, 65–6, 74, 84–8, 107–8, 113–15, 121–7, 129, 134–42, 149–59, 182–5, 225, 255–6; of biotechnology, 14, 55–6, 85–7, 151, 164–73, 183–8, 203; of capitalism, *see* colonization; of Eurocentrism, *sSee* Eurocentrism, critique of; of globalization, *see* colonization; globalization; of positivism, 250, and objectiv-

Index

ity/disembodied rationality, 74, 250, 256, See also science, critique of; of professions/professional schools, 152–9

'culture,' as devalued knowledge, 126, 139; transforming the modernist view of, 263. See also ethnoscience

curriculum, xii, 18, 47, 56, 59, 65, 70–2, 83, 87–91, 98, 110, 112–15, 127–8, 149, 153, 156–9, 186, 196–207, 214–21, 233–6, 270–1, 276–8
- Aboriginal Educational Framework, 233–51
- biology/natural sciences, 186, 271, 276–8; teaching feminist, 70–1, 271, 276–8
- caring/caring rules, in nursing, 160, 214–21; in pharmacy, 156; in professional education, competing with academic capitalism, 157
- change, 72, 87–90, 153, 198, 201–7, 212–13
- contested, 157
- decolonizing. See decolonization, education
- emancipatory, 149, 220–1. See pedagogy; teaching, transformative
- in engineering, 196–200, 203; and gender inclusive curriculum, 199–207
- and environment, 197–8, 204, 276–8
- with equity-relevant skills, 202–3, 205
- exclusionary/androcentric, 128, 197, 199, 270
- feminist, 90, 70–1,112, 190, 196, 203
- forestry, 65–6
- hidden, 98 (pedagogic authority), 150, 155–6 (corporate control), 215
- and Indigenous traditions, 47, 56, 112, 127
- interdisciplinary. See academic, interdisciplinarity
- in medical laboratory science, 89
- in medicine: caring vs curing, 83, 153
- nursing, 149, 210–21. See also caring curriculum
- pharmacy: dominated by basic science, 156–7
- postcolonial, 113, 224; Indigenous framework for, 55–9, 233–6; criteria for, 222–6
- resisting hegemonic curricula, 71–2, 88–90, 153
- in science, 271
- syllabus, 100
- textbook as authority, 98; See also environmental sustainability, teaching, professional education

decolonization/unravelling colonial mentality, 121, 224–5, 229–30; and activism, 56, 225–9; and capitalism, 134, 139; definition, 225; and education, 56, 225, 229–30; and Indigenous knowledges, see knowledges; of scientific method, 230

democracy, 134, 137–46, 182, 186, 275–7; through a corporate lens, 137; and decision-making around wildlife, see wildlife; and depletion of biodiversity, 275–6; and equality, 48, 143, see also equity; and Nature (environmental citizenship), 276; and science, 14, 43, 165, 168–72, 186; and sustainability. See environmental sustainability ; threats to, from biotechnological dominance, 137, 172–3, 186 ; (lack of) in Third World, 138–9

development: economic, 277; as Freedom, 140; 'gap,' 135–6; and globalization/neocolonialism, 135–40; sustainable, 122, see also environmental sustainability, speciesism

discourse/discursive framework. See colonization/colonialism

discrimination, definition of, 227. See also equity, oppression

diversity, 49, 51, 61, 74, 140–3, 199, 207, 229–30; and coalition-building, 143, 199; depletion of bio-, 275; of Indigenous knowledges, 130; of methodologies, 88; scientific, 172; and spirituality, 49–51. See also 'race'

domination, 226, 256. See also Nature, colonization

Earth. See Nature

ecofeminism/environmental/ecological feminism, 2, 8, 61, 71–3, 142, 182, 190, 263–74, 263, 276; and bioregionalism, 270; cyborg, 263, 265; and "environmental citizenship" in praxis, 190, 270; and equity, 273–4; and natural history research, see Nature; and teaching, see teaching

ecology/ecosystems, 272; "of subjects," 276
economics, liberatory/local, 140–1
education:
- Aboriginal Educational Framework, 233–51
- decolonization of, Eurocentric/role in colonization, 121, 125, 224–30
- engineering, *see* engineering, professions
- and environment, 186, 197, 225. *See also* teaching
- and equity, 188. *See also* decolonization, equity
- and evaluation. *See* evaluation
- higher, 149–61, 228–9; and Aboriginal knowledges, 229–31; and need for critical scholarship in professions, 159–61; and funding, 157. *See also* academic, research; and globalization/colonization, 142, 150, 157–9; and Indigenous peoples, 229; as producing human capital, 153; and media, 152; and race. *See* academic, capitalism; race; restructuring; teaching, and women, 111
- of Indigenous peoples, 224
- laboratory, 89
- nursing. *See* professions
- passing on of Indigeonous knowledges, 122
- and performativity, 42
- professional. *See* professional education, capitalism, engineering, medicine
- responsibility for critical teaching, 142; for transforming Aboriginal-Canadian relations, 228
- and science, 42–3, 56, 90, 187
- and spirituality, 48, 56–7
- and women, 47. *See also* women
  See also environmental sustainability and teaching
engineering, 153, 183, 195–207
- As If the world mattered, 198, 205; attrition rate, 206; and "the chilly climate," *see* "chilly climate"; and curriculum, *see* curriculum; education, 15, *see also* professions, engineering; and environment/sustainable development, 153, 195, 200; and equity relevant skills, *see* equity; and feminism, 197–9; and gender differences/issues, *see* gender, women; and language, 203; and pedagogy, and knowledge construction, 197, 198; and professional culture/conduct, 198–200, *see also* Ethics; and society, 195, 198, 202, 203, 205; and teaching, *see* teaching, engineering; and women, *see* women. *See also* professionals, engineering

environmentalism/environmental sustainability/unsustainability, xiii, 12, 15, 57, 61, 66, ; 107, 113, 122, 128–31, 141–2, 151–3, 173, 181–91, 187–91, 204, 225, 272–8; and Canada/US, 151,185–6; criteria for, 61, 212; in the curriculum, *see* curriculum; and the decolonization of Indigenous knowledges, 225; definitions of, 48, 67, 182, 273; and democracy, 141, 182, 188, 276; and ecology/environment, 66, 107, 150–1; and environmental degradation, *see* science ; and feminism/eco-feminism, *see* ecofeminism ; and gardening, 4, 5; global warming, 182; and interdependency, 27–8, 57, 30–2, 122, 142, 244–5, 255–65; interdisciplinarity, 17–18, 43, 113, 187, 189–90; and literacy/illiteracy, 181–91, *see also* literacy/illiteracy/languages; parallels/congruence with equity,1, 4, 8, 17–18, 60–1, 72–3, 76, 113, 141–2, 153, 188–90, 225, 273, 275; personal experiences and, *see* lived experience   ; and race. *See* race; and spirituality, 52, 61, 128, *see also* spirituality; teaching, 59–61, 65–7, 70–7, 113, 153, 197, 269–78; traditional ecological knowledge, 122, 127–31, 270
epistemology/epistemological ; communities, 276; crisis, 65; as dialogism, 32, 81; as engaged relationships, 29–32, 115, *see also* scientific method, 'modest witnessing'; epistemology-based ethics/ethics-based epistemology, *see* ethics; feminist, 60–1, 87, 255–65; and agency, 260–1, and embodied vision, 252–4, 256–65, and knowledge production, 67–9, 108–15, 252–3, 257; Indigenous, 131; positivist, and traditional teaching, 98; postcolonial theory, *see* post-colonialism; practice-based, in professional schools, 153; and spirituality, 47, 52–4, 61. *See also* knowledges; science; scientific method
equality/inequality, 59, 143, 229; from a

critical perspective, 82. *See also* democracy
equity/inequity, xiv, 1, 55–7, 72–3, 90, 108–13, 132, 134–5, 140–1, 149,153–9, 165, 182–3, 188–90, 196–207, 227–8, 273; and academic capitalism, 156; anti- and post-colonialism, 8, 14, 55, 127; in Canada, 188–9; and education, 188–9, 196–7, 226–7, *see also* teaching, transformative; and equity-relevant skills in engineering, *see* curriculum; for faculty, 158–9, 201; and gender, *see* gender; and globalization, 14, 55, 135–9, 154, 159–60, 188–9; and health care organizations, 90, 206; and higher education, 15, 153–9, 197, *see also* higher education; for Indigenous peoples, 132, 139–40, 227, *see also* Indigenous peoples, knowledges; and learning, 57, 90, 197, 207, 213; parallels/congruence with environmentalism, 1, 4, 8, 17, 18, 72–3, 76, 113, 136, ; 141,151, 153, 188–90, 273; and pedagogy, 97–101, 159, 206, *see also* pedagogy; personal experiences and, *see* lived experience; 'race,' *see* racism; in science, 7, 55, 72–3, 165; and spirituality, 57; standards/'excellence' in education and, 154; and teaching, *see* teaching; in Third World, *see* Third World
ethics, 39–43, 59, 66–9, 71–2, 76, 121, 159, 167, 202, 204, 214, 226, 256–65, 273–8; biocentric/environmental, 11, 39–41, 56, 59–60, 65–77, 256–65, 273–8; embodied (caring), 256–60, 265; engineering code of ethics/ethical issues, 202, 204; and an environmental ethics-based epistemology, 274; epistemology-based vs everyday (emergent from practice), 38–41, 43, 56, 69; etiquette/caring-based, 39–41, 69; privacy, 204; in professional schools, 159, 202, 204, 214; research, and Indigenous peoples, 226; and science, 42, 56, 66–8, 71, 167, 256–9, *see* capitalism, biopiracy; and 'universal consideration,' 40–1. *See also* knowledges, ethics-based
ethnoscience (lay knowledge of science), 111, 183–4, 226–7; and biopiracy, *see* biopiracy; and Indigenous knowledges, 130, 183–4, 226–7; and literacy, 183; and multinational corporations, 169, 184. *See also* Indigenous knowledges, public

Eurocentrism/Western science/thought, 123–33, 225–6, 255–65; critique of, 54–6, 65–6, 70, 74, 80, 83, 87–8, 107–8, 113–15, 123–32, 151–2, 164–7, 183–4, 255–65; appropriating Indigenous knowledges, 123–127, 132, 183–184; originating through exposure to science, 169; and pedagogy, 17, 57, 113, *see* also pedagogy and the professionalization of science, 107–108. *See also* colonization, critique of Eurocentrism, scientific method
evaluation, educational, 15–16, 98–100, 155, 210–21; in caring paradigm in nursing ("velvet hammer"), 210–21; by faculty, 214–21; of faculty, 155; formative vs summative, 214; and professional standards, 210–21; student distrust of, 216; traditional, 215, 218; and vulnerability, 216–20. *See also* professional education

faculty. *See* equity, for faculty; evaluation, of, by faculty
feminisms, 69, 70, 142, 182, 273–4. *See also* academia; curriculum; ecofeminism; engineering; epistemology; pedagogy; science, feminist and environmental sustainability; scientific method, feminist; knowledges, construction; critique of Western science; academic, research
Franklin, Ursula, 6, 151, 198
Friere, Paolo, 9, 73, 81, 213

gender, 12, 72–3, 90, 96, 107, 109, 111, 115, 134, 152, 154, 156–7, 184, 195, 199, 201, 205; bias: in engineering, 196–201, 204, in forestry, 66, in medical laboratory technology, 82, 90, in pharmacy education, 156, in science, 72, 107, 109, 111, 152, 164–5, 184; and environment, 113, 115; inclusive language, *see* language; issues/differences, in engineering, 193–4, 195, 197, 199, 200, 202–7; and race, *see* race; in teaching, *see* power relations. *See also* women; lived experience
genetically modified organisms. *See* biotechnology
global warming. *See* environmental sustainability
globalization, 13, 55–6, 135–45, 150–2,

154–5, 159, 188–9, 198, 230–1, 275, 278; alternatives to, 55–6, 139–43, *see also* post-colonialism; and capitalism, 55, 135–41, 151, 188–9; critique of, 7, 55, 135–41,150–1, 188–9, 278; and dehumanizing effects of restructuring, 84, 158–9; and development, 16, 55–6, 121, 139, 151, 226, 231, 275, *see also* development; Indigenous knowledges; and engineering, 198; and environmental unsustainability, 189; and neo-liberal paradigm, 135–9, 188–9; and speciesism, 17, 151, 275; and Third World countries, 135–8, 151; and WTO, 138

government/the state, 227 (responsibilities around discrimination); Anishinaabe, 246–8; and environment, 277

Haraway, Donna, 7, 68–9, 87, 152, 256–65, 273
Harding, Sandra, 7, 68–9, 72, 75, 152, 165, 167, 186
higher education. *See* education
homophobia. *See* oppression
hope/existential angst: and activism, 161, 191, *see also* activism; about the environment, 5, 65, 75, 77, 170
Hubbard, Ruth, 85, 152

imperialism. *See* colonization/colonialism
Indigenous/Aboriginal knowledges 6, 11, 12–17, 47–57, 59, 69–70, 111–14, 121–32, 151, 224–31, 233–51; appropriation of, *see* colonization; balance, 240, 245, 249; centring/destruction/devaluation, 50, 54–9, 61, 121–33, 167, 226; and decolonization, *see* decolonization, colonization; definitions of, 122, 128–32; diversity of, 129–30; and education, 121, 127, 224–5, 229–31; ethical basis of knowledge, 40, 56, 69–70, 121; of grandfathers, 248; and holism, 128–9, 131–2; and the land/environment, 28–9, 53–4, 59–60, 129, 151; and mapped relationships, 29; of mothers and grandmothers, 48, 50–3, 60, 247–9; passed from generation to generation, 122, 129, 233–51; and pedagogy, *see* pedagogy; and pharmaceutical companies, 123, 221, *see also* biopiracy; renaissance and decolonization, 225; and responsibility, *see* responsibility; science and spirituality, 54, 56, 131–2; and song, *see* spirituality; and spirituality. *See* spirituality; technology, 128; and transcultural coalitions, 51, 54, 143, 145, 224; wisdom to interrogate oppression, 50–1, 57–9, 121; and women, 114

Indigenous peoples, 121–32, 224–31, 233–51; Anishinaabe Clan system, 246–7; and Canada (RCAP), 126, 228–31, 233–4; and colonization, 228, *see also* colonization, anti-colonialism; community, as basis for learning, 239; and curriculum, *see* curriculum, post-colonial framework; and education, 224, 229, 234; Elders, 233–9, 247–51; and Eurocentrism, 54–5, 122–5, 225–6. *See also* critique; and genocide, 228; and global racism, *see* racism; and higher education, 224–25; and knowledges, *see* Indigenous knowledges, knowledges, Aboriginal; spirituality; learning, 234; and ; and relationships with each other, environment, *see* relationships; and student-teacher relationships, *see* student-teacher relationships; and United Nations, 121, 131, 227–30. *See also* Indigenous knowledges; relationships between Indigenous peoples

industry, impact on nature, 272, *see* capitalism; knowledges, Indigenous; academic, capitalism; curriculum, pharmacy; science, 84, 151
inter-/multidisciplinarity. *See* academic, research, teaching, professions
intolerance/tolerance, 226, 229. *See* Eurocentrism, oppression;

knowledges, 50–1, 54, 56, 65, 67–70, 83, 87, 100, 111–14, 121–31, 151–2, 156–60, 164, 184–91, 224–9, 240–51, 256–62, 271, 274; Aboriginal, *see* Indigenous knowledges; anti- and post-colonial, 13, 14, 56, 113; As If the world mattered, 65, 115, 185–91; biotechnological, *see* biotechnology; bioregional, 271; community-based, 40, 112–13, 233–51; construction/production of, 12, 18, 66, 69, 70, 87, 100, 130–1,152, 159, 164, 194, 261–2; "through death"/violence, 87, 256, 258–62; devaluing/valuing of, *see* academia; diffusion of, and power, 226; and emotions, *see* teaching; in engineering,

Index   287

153, 194, 198; and epistemology, *see* epistemology ; ethics, *see* ethics; Eurocentric/Western, 13, 40, 50, 54–6, 66, 69, 82, 113, 122–7, 130–31, 134, and primacy of teacher as "more objective," 96–8; feminist, 7, 16–17, 50–3, 60, 108, 112, 152, 165, 190, 249–61, 256–65; neoliberal, 70, *see also* science, feminist, ecofeminism, Indigenous, *see* Indigenous knowledges; in medical laboratory technology, 85–9; production, 259; racialized, *see* race; and respect for earth, 122, 190–1, 244–5, *see also* scientific method, 'modest witnessing'; and responsibility, *see* responsibility; situated/contextualized, 68, 164–5, 256, 259–62, 275, *see also* scientific method, objectivity, critique, of science; and spirituality, 11, 56, 70, 245; subjugated/suppressed, 1, 12, 13, 17, 67, 108, 114, 123–5, 156, 159–60, 167, 229, *see also* Indigenous knowledges, writing; traditional ecological (TEK), 129, 244–5; ways of knowing, *see* epistemology; Western, *see* Eurocentric/Western, critique, science. *See also* science

languages/literacy/illiteracy, 171, 181–91, 197–8, 200, 203, 205, 230; definitions of literacy, 181–4, 187, 190; in engineering, 197, 200; and equity, 187; gender-inclusive, 203, 205; and Indigenous knowledges, 122, 126–31, 230; of the public about science, *see also* public; of social scientists about science, 182–6; of scientists about the constructed nature of knowledge, 164, 185–6; and transformative teaching, 102
Lewontin, Richard, 27, 261
life/growth. *See* Nature, trees/plants, wildlife
lived experience: as access to Indigenous knowledges, 130, 233–51; of animals/bioregion and community planning, 271, 277; and bifurcation in knowledge, 80, 199–200; and environmentalism, 4, 5, 10–12, 25–7, 29–33, 35–6, 41, 65–6, 77, 107, 130, 166, 182, 260, 276–8; and equity, 4, 5, 7, 11–12, 16, 90, 107, 115, 134, 149–50, 165, 211–12, 260, becoming 'visible,' 80–1, *see also* writing, women's, integrated into science; of failure, 210–12; of scientists, 110, 115, 260; and spirituality, 52; and transformative teaching, 73–4, 95, 102, 216–17, 276–8; use as pedagogy, 73–4, 98–9, 115, 276–8

media, and globalization, 136, 142, 145; coverage of educational issues, *see* education, higher
medicine: and medical laboratory technology, 82–6; sacredness of, 245. *See also* professions, medicine; medical laboratory technology
minorities. *See* race, oppression

Naess, Arne, 9, 42
Nature/Mother Earth/Earth/Turtle Island, 122, 246, 269–78; and agency/intentionality/movement, 261; devaluation of, 274–5; domination of, 66, 150–1, 166, 256, 274; as 'solution' to environmental crisis, 171–3 ; as healing, 149–50; Indigenous relationship to, 122, 270; industrial models of, 151, 170; learning about, 244–5, 271–2; -as-miracle, 275; as a model for human constructions, 67; natural history, 114, 271–3, *see also* knowledges, subjugated, colonization of; protection of, 122; respect for, 76, 191, 244–6, 255–65, 270, 274; and spirituality, 53, 57, 59–60, 244–6; transforming the modernist view of, 263, 269–78. *See also* respect, wildlife, environmental sustainability
nuclear issues, xiii–xiv, 42, 56, 185–6, 189
nursing: and responding "as a nurse," 212, 215, 216; and curriculum, *see* curriculum; and education, 206; and evaluation, *see* evaluation; and power, 213–14; and professional socialization, 210–13; and student teacher relationships, 210–18; and vulnerability, *see* relationships. *See also* pedagogy, human science paradigm

oppressions, 154, 273, 206, 218, 226, 260, 273; ableism, 260, 273; and academic evaluation, 218; homophobia, addressed in the engineering curriculum, 204; intersections of, 134, 273; and medical laboratory technology, 81–3, 84–6; speciesism, *see* speciesism. *See*

also colonization; equity; Eurocentrism; gender; race; Third World
Orr, David, 9, 67, 74

peace/militarism/weapons. *See* activism, for peace; nuclear issues
pedagogy, 9, 15–16, 57–60, 73–7, 87–9, 95–103, 112, 153, 196–7, 203, 210–15, 277; anti-colonial/anti-racist, 58, 95–8, 191, *see also* 'race'; and capitalism, 96 ; caring vs distancing, 210–15; classroom, 33, 57–60, 95; community, 233–51, 277; consciousness-raising/critical, 9–12, 58, 73–4, 95–8, 87–9, 99, *see also* teaching, transformative; critical pedagogy, 95–102, *see also* teaching as activism; definitions of (including knowledge production), 95–9; and developing students' spiritual dimensions, 57–60; engineering, 193; environmental, 277, *see also* teaching; feminist vs androcentric, xiv, 9, 12, 15–17, 94, 112, 197, 203; and higher education, 17, 57, *see also* professional education; holistic, in professional schools, 153, 201–7; and inculcation of habitus, 98–9; and mathematics, 87–9; nursing, 210–21; and power relations, 56–7, 59, 95, 96–101, 210–13; and professional education, *see* professional education; and reflective processes, 59; rule-based vs relational, 196, 210; spiritual (circle, drumming, teaching from Elders, meditation, singing, prayer, cleansing), 58–60; traditional, 98–9, 213; transformative, *see* teaching, transformative; and Whiteness, 17, 18
personal experience. *See* lived experience
plants. *See* trees
policy: development, 140; educational, 115; environmental, 152, 273; neocolonial, 135; public, and genetically-modified organisms, 167–72; and science, 42–3, 67; in wildlife management, 38; science, 110, 115, 152, *see also* public, democracy
post-colonialism; and curriculum, *see* curriculum; and Indigenous knowledges/people, 121–3, 126, 224–30; post-colonial theory, 121, 112–13, 227–30; and critique of Eurocentrism, *see* colonization, racism. *See also* education, Aboriginal Educational Framework; knowledges; Indigenous knowledges
power relations, 88–91, 95–8, 98–103, 107–11, 134–43, 154–61, 188, 210–21, 226, 256–60, 255–65, 273, 278; colonial. *See* colonization; and control of knowledge, 226, 256, *see also* Eurocentrism, Indigenous knowledges, knowledges; discussed in sustainable science course, 74; examined in curriculum, 88, 159; in health care organizations, 90; in professions/professional education, 154–61, 210–21; reproduced through teaching, 13, 95–101, 159, 210–14; in science (relations of ruling) 13, 57, 85–7, 107–11, 165, 167, 255–65; as vision, 256–65. *See also* colonization; Eurocentrism; gender; "race"
praxis. *See* teaching, transformative; professions, practicum
professionalization of science. *See* critique of science
professions/professional education, 12, 65–6, 80–92, 149–61, 195–207, 210–21; architecture, 153, 155; and corruption, 155; curriculum, *see* curriculum; dentistry, 157; deskilling, 158; engineering, 153, 195–207, *see also* women, in engineering; and failure, 210–14, *see* evaluation; forestry, 65, 150, 156; and hierarchy, 82–6; as human capital, 153–4; and inequity, 153–4, 200, *see also* gender; and lack of critical scholarship, 152, 159; law, 157, 158; medical laboratory technology, 80–92; medicine, 153–4, 160; and multidisciplinary teams, 199; nursing, 210–21; optometry, 157; pedagogy, 213–14; and pharmaceutical industry, 14–15, 149, 151; pharmacy, 156–9; and power relations, *see* power relations; practicum/professional practice courses/praxis, 203, 205, 210, disjuncture and professional associations, 200, 205; and professional socialization, 210; and science/biology. *See* science; and standards. *See* evaluation; between theory and practice, 89–91, 157, 160, 202, *See also* capitalism, and professions; curriculum; education, professional
public, awareness/literacy about professions/science/technology, 83, 111, 171–2, 182, 186–7, 265; and passive

resistance, 8, 171 (Europeans to GE foods), 182; debate/distrust of biotechnology, 171, 186, 192n; 'good' and democracy, 276, argument to control knowledge, 226, *see* Indigenous knowledges, centring/devaluation of; interest and professions, 154; misunderstanding of science, 171. *See also* ethnoscience, languages/literacy, democracy

'race'/racism/ethnoracism/anti-racism, 59, 82, 88, 95, 111–12, 114, 123, 126–7, 134–6, 151, 153–4, 157–8, 160, 204, 227, 243, 252n; in Anishinaabe teachings, 243–4; anti-racism/anti-racist pedagogy, 59, 76, 95; and biopiracy, 123, 226–7, *see also* Indigenous knowledges; definition of 'race'/racism, 6–7, (from Memmi), 126; and environment, 153, 190, Eurocentrism as racism, 125, 135; and gender, 135–6, 158, *see also* women, Third World; global racism, 123, 167, 227; and knowledge production, 12, 123–5, 157–61; post-colonialism and, 113–14, 125–6; in professional education, 154, 157–61; in science, 72, 82, 88, 110–11, 151; and gender, 112, 114
reductionism, scientific/market, 83, 87, 173. *See also* biotechnology; capitalism; science; scientific method
reflective processes, 68, 96, 102, 149–50, 165, 167–8, 216, 242, 270. *See also* pedagogy; teaching, transformative
relations of ruling. *See* power relations
relationships: between Canadians and Aboriginal peoples, 126–7; curricular, 88–90, 206; between four elements, 51, 246; between humans and other life, 36–9, 44, 59–60, 68, 122, 128–9, 191, 244–5, ; 255–65, 270–1, 274–5, *see also* wildlife, speciesism, trees; between (Indigenous) peoples, 122, 241–9; between plants and environment, 27–30; between scientists and non-humans, 262; between scientists and the public, 168, *see also* public; split by technology, 48, 81, 256; between subject and object, 257–62; between teacher and learner. *See* student-teacher relationships. *See also* power relations
research, 83, 112, 115, 156–7, 187, 225–6; academic, 1, 65–6, 108, 110, 112, 151–2, 154–5, 157–8, 164–5, 169; Indigenous, 225; interdisciplinary, 88, 187; in medical laboratory technology, 88 ; and teaching, 112, 115, 156–7. *See also* biotechnology; capitalism; science, biologists; scientific method/research
respect: of Canadians for Indigenous peoples, 127, 228–9; between cultures, 141, 229 ; for diversity resulting from classroom critique of colonialism, 75; between Indigenous peoples, 49, 122. *See also* knowledges, respect; Nature; relationships; responsibility; trees; wildlife
responsibility (caring/love), 237, 251, 259–65, 274; based on an ethic of kinship, 59, 69–70, 274; of Canadians/the state towards Aboriginal peoples, 127, 225–31; for encouraging spirituality in education, 57–8; in knowledge production, 69–71, 259–65, *see also* scientific method, strong objectivity; societal, in engineering, 198; to students, 215, 236–7 (Aboriginal); for utilizing knowledge in a truthful way (Anishinaabe), 240, 251
restructuring. *See* globalization, restructuring

science/biology, 3, 37–9, 42–3, 48, 55, 66–73, 83–9, 107–13, 122, 129–30, 151, 164–8, 173, 183–91, 210, 255–65, 274–5; amateurs and devaluation of fields, 108, 111, 114, 184; biologists/scientists/research personnel, 37–9, 42, 66, 107–11, 114–15, 130, 157–8, 166, 168, 184–6, 229, 256, 260, 262, 264, and failure to understand Indigenous knowledges, 122, 129–31, social sciences, 186, *see also* women, in science; bioregional, 271; and capitalism, 151; careers for Whites vs men of colour, 157–8, for women vs men, 107–8, 111–12, 114, 156–7, 200; critique of, 183–5, 255–6, *see also* critique of Eurocentrism, ethnoscience; definition of, 112, 183–4, *see also* feminist science, scientific method; and democracy, *see* democracy; and devaluation of physical work, 85; and dualism/integration, 43, 81, 112–13, 274–5; and efficiency, *see* scientific method; environment impact, 6, 42,

65–7, 182; and environmental degradation/toxicity, 14, 65–7, 150, 272–3; and (in)equity, 55, 108, 165; ethics, *see* ethics; ethnoscience, *see* ethnoscience, Indigenous knowledges; 'exact' measurement in, *see* scientific method; feminist, 152, 264, *see also* scientific method, feminist; and government, 111; human (nursing), 210–14; illiteracy and social scientists, 17, 181–3, 186; and public misunderstanding of science, *see also* languages/literacy/illiteracy, public ; Indigenous, 54–7, 69–70, 112, 130; and the laboratory/experimentation, *see* scientific method; and (scientific) method, *see* scientific method ; and objectivity, *see* scientific method; and performativity, *see* scientific method; post-colonial and anti-colonial, 17, 48, 57, 113, 152; and power relations, *see* power relations; and privileging of rationality, 86, 172, 184, 256 ; and race, *see* 'race'; reform, 86; research, *see* academic, research, capitalism, research; and sexism, *see* science, feminist, gender, women, in science; and spirituality, *see* spirituality, knowledges, Indigenous, subversive/insurgent, 16, 17, 272; sustainable, 69–70, 173, *see also* environmental sustainability, teaching; as not value-free, 73, 83, 86–8, 164, 167, *see also* scientific method, objectivity; and women, 165–7, 168–70

scientific research/method/theory, 66, 80, 84, 86, 168, 171, 186–90, 226, 230, 255–65; and agency/movement of non-humans, 257–62; and efficiency, 84; 'exact' measurement in, 88; feminist, 7, 66–71, 107, 165, 256–60; and "a feeling for the organism," 30–1, 258; vs Indigenous search for truth, 226, 245, 247, 250; and the laboratory/experimentation, 259–63; limits of, 37–9, 41, 70, 164, 167–8, 186–90; and "modest witnessing" vs "gaze," 257–60; and objectivity/disembodiment, 37–9, 66, 83–6, 164, 255–60, strong objectivity (environmental responsibility), 68, 72, 256, 259–65; and 'neutral' observation/distancing vs embodiment, 255–60; and performativity, 42; and violence, 256, 258–62; and vision, transubstantiation into touch,

255–65, *see also* epistemology; and visualization technology, 255–65. *See also* decolonization

sexism, 200, 204. *See also* gender, women, academia

Shiva, Vandana, 8, 136, 151, 158, 184

Smith, Linda Tuhiwai, 13, 50, 131, 224–5, 229

speciesism, definition, 4; and anti-speciesism as inclusiveness, 76, 271; critiqued in an ecofeminist course, 73, 76; vs an ethics of respect, *see* ethics, universal consideration; and globalization, 17; as lack of respect for wildlife, 40–1, 270; and literature, 8–9, 68; and science, 72; and social science, 15, 68. *See also* nature, scientific method

spirituality, 10–11, 16–17, 47–61, 70, 128–9, 233–51, 277; and activism. *See* activism; and bioregionalism, 277; and community/connection, 48, 52, 57–8, 60–1; and Creation/Creator, 233–51, 275; and Creation/land/environment, 48, 52–4, 61, 128–9; definition as the "science of our soul," 48, 51, 56; as embracing diverse (Indigenous) cultures, 49, 51–2, 54; everyday vs academic concepts, 53; as experience of being, 52; and Indigenous knowledges, 50–60, 128–9; as intuitive/"inner," 52, 54; and Nature, 275; and religion, 52; role of animals, 245; sacredness of individual/responsibilities/life, 236–45; separation from the hard sciences, 40, 57; seven laws of the Anishinaabe, 240; spirits of the medicine plants, 28, 31, 244–5; and song, 29, 128, 237, 245; sustainability, *see* environmental sustainability; and transformation of consciousness, 53; Western vs Indigenous knowledges, 69–70, 129; and wisdom, 52. *See also* curriculum, postcolonial framework; education; pedagogy, environmental sustainability; responsibility

student-teacher relationships, 96–101, 210–21, 236; in Aboriginal education, 57–60, 234–51; "authentic presencing," 220–1; in nursing/failure, 210–21; and vulnerability, 212–13, 216–21. *See also* evaluation; curriculum, caring; pedagogy, caring; teaching/learning

Index 291

sustainability. *See* environmental sustainability, development

teaching/learning, 95–103, 152, 159, 196–7, 206, 210–21, 233–51, 270, 274; as activism, 3–4, 16, 33, 47, 54–61, 65–77, 95–103, 142, 152, 159–60, 203, 213; ; 'banking' vs caring model, 210–18; classroom, 33, 57–9, 75–7, 95–102, 197, 203, 205; location of, 59–60; clinical/experiential, 210–18; critical. *See* transformative, critique; deficit model, 210, 217; and ecofeminism, 73, 264, 276–8; by Elders, 233–9; and emotions, 58–9, 75, 217, 271, 274; engineering, 201–7; environmental sustainability, *see* environmental sustainability; and equity, 33, 57, 156–9, 276; evaluation/failure, *see* evaluation; and feminist pedagogy, *see* pedagogy; and gender differences, 196–203; habitus, 95–100; and higher education, 98, 101–2, *see also* professional education; interdisciplinary, 89, 112, 115, 202; literature, 8–9, 95–103; mentoring/role models, 206, 213; and personal experiences, *see* lived experience; and power relations, *see* power relations, pedagogy and power relations; in professional education, *see* professions; and relations of ruling, *see* power relations; and research, *see* research; and spirituality, 57–8; and standpoint, 98; student-teacher relationships, *see* student-teacher relationships; sustainable science, *see* environmental sustainability; transformative, 12, 16, 48, 60, 95–103, 201–7, 211, 216, 276, *see also* pedagogy; wand, 235–51. *See also* education; evaluation; environmental sustainability; pedagogy

technology: vs science, 86; transfer, 137; transforming the modernist view of, 263–4; visualization, limits of, 255–65. *See* biotechnology; women

Third World, 134–46. *See also* capitalism, exploiting Third World

trans-/multinational corporations. *See* capitalism

trees/forests/plants, 4–5, 25–31, 65–6, 59–60, 65–6, 76, 18, 271; and gardening, 5–6; possessing agency/movement, 261; sacred in Anishinaabe teaching, 244–5. *See also* Nature, speciesism

United Nations, and Indigenous peoples. *See* Indigenous peoples
universality. *See* critique, of Eurocentrism
university/universities. *See* higher education

vision. *See* scientific method, epistemology

war-making. *See* activism, for peace
ways of knowing, and epistemology, 13, 183, 196–200; participatory, 70, 198. *See also* epistemology; scientific method
wildlife, and agency/movement, 261, 278; accepted by children, and biocentric ethics, 274–6; and curiosity/knowledge of, 10–11, 35–6, 271, 275–6, 278; and human interrelationships, *see* relationships; management, 36–9, 41; ; porcupines, 269–70, 278; and public decision-making, 38; respect for, 40–1, 244–5, 270, *see* Nature; as 'useless,'/pests, 270, *see also* speciesism; wolves, 35–43
wisdom. *See* knowledges, spirituality
women: in academia, 154–8, 199–202; and engineering, 15, 152, 168–70, 195–207; and forestry, 66; and globalization, 135; and learning, 196–8; and medical laboratory technology, 82, 86; National Conference for, in Science, Engineering and Technology, 108; in ornithology, 108–9; and patriarchy/androcentrism, 134, 197; in pharmacy education, 156–9; as role models, 201; in science and technology, 107–15, 168–70; course, 110, 112–15; rejecting feminist epistemologies, 87, 206; unaware of androcentric language, 203; in the Third World, 134–6. *See also* gender
writing and (in)visibility, 80, 86, 109, 114, 256. *See also* knowledges, subjugated